Lecture Notes in Economics and Mathematical Systems 637

For further volumes:
http://www.springer.com/series/300

Radu Ioan Boţ

Conjugate Duality in Convex Optimization

Springer

Dr.rer.nat.habil. Radu Ioan Boţ
Chemnitz University of Technology
Faculty of Mathematics
Reichenhainer Str. 39
09126 Chemnitz
Germany
radu.bot@mathematik.tu-chemnitz.de

ISSN 0075-8442
ISBN 978-3-642-04899-9 e-ISBN 978-3-642-04900-2
DOI 10.1007/978-3-642-04900-2
Springer Heidelberg Dordrecht London New York

Library of Congress Control Number: 2009943057

Cover design: SPi Publisher Services

Printed on acid-free paper

Springer is part of Springer Science+Business Media (www.springer.com)

To my family

Preface

The results presented in this book originate from the last decade research work of the author in the field of duality theory in convex optimization. The reputation of duality in the optimization theory comes mainly from the major role that it plays in formulating necessary and sufficient optimality conditions and, consequently, in generating different algorithmic approaches for solving mathematical programming problems. The investigations made in this work prove the importance of the duality theory beyond these aspects and emphasize its strong connections with different topics in convex analysis, nonlinear analysis, functional analysis and in the theory of monotone operators.

The first part of the book brings to the attention of the reader the perturbation approach as a fundamental tool for developing the so-called conjugate duality theory. The classical Lagrange and Fenchel duality approaches are particular instances of this general concept. More than that, the generalized interior point regularity conditions stated in the past for the two mentioned situations turn out to be particularizations of the ones given in this general setting. In our investigations, the perturbation approach represents the starting point for deriving new duality concepts for several classes of convex optimization problems. Moreover, via this approach, generalized Moreau–Rockafellar formulae are provided and, in connection with them, a new class of regularity conditions, called closedness-type conditions, for both stable strong duality and strong duality is introduced. By stable strong duality we understand the situation in which strong duality still holds whenever perturbing the objective function of the primal problem with a linear continuous functional. The closedness-type conditions constitute a class of regularity conditions recently introduced in the literature. They experience at present an increasing interest in the optimization community, as they are widely applicable than the generalized interior point ones, a fact that we also point out in this work.

We employ the conjugate duality in establishing biconjugate formulae for different classes of convex functions and, in the special case of Fenchel duality, we offer some deep insights into the existing relations between the notions strong and stable strong duality. Moreover, we enlarge the class of generalized interior point regularity conditions given for both Fenchel and Lagrange duality approaches by formulating corresponding sufficient conditions expressed via the quasi-interior and quasi-relative interior.

The convex analysis and, especially, the duality theory have surprisingly found in the last years applications in rediscovering classical results and also in giving new powerful ones in the field of monotone operators. Among others, we provide a regularity condition of closedness-type for the maximality of the sum of two maximal monotone operators in reflexive Banach spaces, which proves to be weaker than all the other generalized interior point conditions introduced in the literature with the same purpose.

I express my thanks to Gert Wanka for his incessant support and for giving me the possibility to do my research in such a nice academic environment like Chemnitz is offering. Thanks also to Ioana Costantea, Ernö Robert Csetnek, Sorin-Mihai Grad, Andre Heinrich, Ioan Bogdan Hodrea, Altangerel Lkhamsuren, Nicole Lorenz, Andreea Moldovan and Emese Tünde Vargyas, former and current members of the research group at the university in Chemnitz, with whom I share not only an intense scientific collaboration but also a nice friendship. I am indebted to Ernö Robert Csetnek, Anca Grad and Sorin-Mihai Grad for reading preliminary versions of this work and for their suggestions and remarks.

I am grateful to my parents and to my sister whose encouragement and support I felt all the time even if they are far away. Finally, I want to thank my wife Nina and my daughter Cassandra Maria for their love, patience and understanding.

Chemnitz, Germany,
August, 2009

Radu Ioan Boţ

Contents

Symbols and Notations

Sets, Elements and Relations

X^*	Topological dual space of X
$\omega(X, X^*)$	Weak topology on X induced by X^*
$\omega(X^*, X)$	Weak* topology on X^* induced by X
$\widehat{x}$	The canonical image in X^{**} of the element $x \in X$
B_X	The unit ball of the normed space X
$\operatorname{int}(U)$	Interior of the set U
$\operatorname{cl}(U)$	Closure of the set U
$\operatorname{lin}(U)$	Linear hull of the set U
$\operatorname{aff}(U)$	Affine hull of the set U
$\operatorname{co}(U)$	Convex hull of the set U
$\operatorname{cone}(U)$	Conic hull of the set U
$\operatorname{coneco}(U)$	Convex conic hull of the set U
$\operatorname{core}(U)$	Algebraic interior of the set U
$\operatorname{icr}(U)$	Intrinsic core of the set U
$\operatorname{sqri}(U)$	Strong quasi-relative interior of the set U
$\operatorname{ri}(U)$	Relative interior of the set U
$\operatorname{qri}(U)$	Quasi-relative interior of the set U
$\operatorname{qi}(U)$	Quasi interior of the set U
Δ_{X^m}	The set $\{(x, \ldots, x) \in X^m : x \in X\}$
$N_U^{\varepsilon}(x)$	ε-normal set to the set U at x
$N_U(x)$	Normal cone to the set U at x
$T_U(x)$	Bouligand tangent cone to the set U at x
$\leq_C$	The partial ordering induced by the convex cone C
∞_C	A greatest element with respect to the ordering cone C attached to a space
$Z^\bullet$	The space Z to which the element ∞_C is added
C^*	Dual cone of the cone C
$\mathbb{R}^T$	The space of all functions $z : T \to \mathbb{R}$
$\Delta_{\mathbb{R}^T}$	The set of the constant functions $z \in \mathbb{R}^T$

Functions and Operators

$\langle x^*, x\rangle$	The value of $x^* \in X^*$ at $x \in X$
$\operatorname{dom} f$	Domain of the (vector) function f
f^*	Conjugate function of the function f
f_U^*	Conjugate function of the function f with respect to the set U
f^{**}	Biconjugate function of the function f
$\operatorname{epi} f$	Epigraph of the function f
$\widehat{\operatorname{epi}} f$	The symmetric of $\operatorname{epi} f$ with respect to the x-axis
$\operatorname{cl} f$	Lower semicontinuous hull of the function f
$\operatorname{co} f$	Convex hull of the function f
$\partial_\varepsilon f(x)$	ε-subdifferential of the function f at $x \in X$
$\partial f(x)$	Subdifferential of the function f at $x \in X$
δ_U	Indicator function of the set U
σ_U	Support function of the set U
id_X	Identity function on X
Pr_Y	Projection operator of the space $X \times Y$ on Y
$f_1 \Box \ldots \Box f_m$	Infimal convolution of the functions $f_i, i = 1, \ldots, m$
$f \vee g$	Pointwise supremum of the functions f and g
A^*	Adjoint operator of the linear continuous operator A
Af	Infimal function of the function f through the linear continuous operator A
$\operatorname{epi}_C g$	C-epigraph of the vector function g
$(z^* g)$	The function $\langle z^*, g(\cdot)\rangle$, where g is a vector function and $z^* \in C^*$
$S : X \rightrightarrows X^*$	Set-valued operator on X with values in X^*
$G(S)$	Graph of the operator S
$D(S)$	Domain of the operator S
$R(S)$	Range of the operator S
φ_S	The Fitzpatrick function associated to the operator S
ψ_S	The Penot function associated to the operator S
h_S	Representative function associated to the operator S
$S_{h_S}^\varepsilon$	Enlargement of the operator S defined via the representative function h_S

Generic Notations

$(P^{\cdots})$	Primal optimization problem
$v(P^{\cdots})$	The optimal objective value of the problem $(P^{\cdots})$
$(D^{\cdots})$	Dual optimization problem
$\Phi^{\cdots}$	Perturbation function
$(RC^{\cdots})$	Regularity condition

Introduction

The purpose of this work is to present the state of art but also recent advances and applications in the theory of conjugate duality in the convex analysis and convex optimization and, beyond them, in nonlinear analysis, functional analysis and the theory of monotone operators. Unless otherwise specified, the content of this manuscript is represented by the contributions of the author (along with his co-authors) to this field. The diversity of the topics considered here constitutes an evidence for the important role which the duality theory plays in the different areas enumerated above. The work is divided into six chapters with a total of 26 sections.

Chapter I. The first chapter is dedicated to the formulation of conjugate dual problems by means of a general perturbation approach for different classes of primal problems which occur in convex programming. Considering a so-called *perturbation function* $\Phi : X \times Y \to \overline{\mathbb{R}}$, where X and Y are supposed to be separated locally convex spaces, one can attach to the optimization problem

$$(PG) \quad \inf_{x \in X} \Phi(x, 0)$$

the following dual problem

$$(DG) \quad \sup_{y^* \in Y^*} \{-\Phi^*(0, y^*)\},$$

where $\Phi^* : X^* \times Y^* \to \overline{\mathbb{R}}$ is the conjugate function of Φ, while X^* and Y^* are the topological dual spaces of X and Y, respectively. Some facts connected to this approach, which has been well-described in the monographs due to Ekeland and Temam [67], Rockafellar [116] and Zălinescu [127] are recalled in the first section along with some considerations concerning the existence of weak and strong duality for the primal–dual pair of optimization problems. We call *strong duality* the situation when the optimal objective values of the primal and dual coincide and the dual problem has an optimal solution. In this section, some *generalized interior point regularity conditions* expressed via the perturbation function Φ are also resumed and in Theorem 1.7 a strong duality result is given by collecting the corresponding results from [67, 113, 116, 127]. In Section 2 the general approach is

R.I. Boţ, *Conjugate Duality in Convex Optimization*, Lecture Notes in Economics and Mathematical Systems 637, DOI 10.1007/978-3-642-04900-2_0,

particularized to the case of the primal optimization problem

$$(P^A) \quad \inf_{x \in X} \{f(x) + (g \circ A)(x)\},$$

where $f : X \to \overline{\mathbb{R}}$ and $g : Y \to \overline{\mathbb{R}}$ are proper and convex functions and $A : X \to Y$ is a linear continuous operator fulfilling $A(\operatorname{dom} f) \cap \operatorname{dom} g \neq \emptyset$. By considering an appropriate *perturbation function* one obtains as conjugate dual of (P^A) its Fenchel dual problem

$$(D^A) \quad \sup_{y^* \in Y^*} \{-f^*(-A^* y^*) - g^*(y^*)\}.$$

Generalized interior point regularity conditions and strong duality theorems for this primal–dual pair and for some particular instances of it are delivered by specializing the statements provided in the previous section for (PG)–(DG). In this way, we rediscover the regularity conditions and the corresponding strong duality assertions stated by Rockafellar in [112, 113], Rodrigues and Simons in [118] and Attouch and Brézis in [2], respectively. In the next section, similar investigations are made for the optimization problem with geometric and cone constraints

$$(P^C) \ \inf_{x \in \mathcal{A}} f(x),$$
$$\mathcal{A} = \{x \in S : g(x) \in -C\}$$

with X and Z separated locally convex spaces, where Z is partially ordered by the nonempty convex cone $C \subseteq Z$, $S \subseteq X$ a nonempty convex set, $f : X \to \overline{\mathbb{R}}$ a proper and convex function and $g : X \to Z^\bullet = Z \cup \{\infty_C\}$ a proper and C-convex function fulfilling $\operatorname{dom} f \cap S \cap g^{-1}(-C) \neq \emptyset$. By means of different perturbation functions, we introduce three conjugate dual problems to (P^C), namely the so-called *Lagrange*, *Fenchel* and *Fenchel–Lagrange* duals, the latter being a combination of the other two. A general scheme illustrating the relations between the optimal objective values of the three duals is provided (see also [14, 39, 124]). By considering along the convexity assumptions some topological additional hypotheses for the sets and functions involved, we also state generalized interior point regularity conditions and strong duality theorems for these three primal–dual pairs. One of the condition we obtain in this way in connection to Lagrange duality is the classical *Slater constraint qualification*, but also a number of regularity conditions in Fréchet spaces is delivered for each of the three duality schemes treated here. It is worth noticing that, again with respect to Lagrange duality, some classical regularity conditions are rediscovered as are the ones formulated by Rockafellar in [112] in Banach spaces or by Jeyakumar and Wolkowicz in [87] for linear programming problems. In the fourth section we work with the composed convex optimization primal problem

$$(P^{CC}) \quad \inf_{x \in X} \{f(x) + g \circ h(x)\}.$$

Here X and Z are separated locally convex spaces, where Z is partially ordered by the nonempty convex cone $C \subseteq Z$, $f : X \to \overline{\mathbb{R}}$ is a proper and convex function, $g : Z \to \overline{\mathbb{R}}$ a proper convex and C-increasing function fulfilling, by convention, $g(\infty_C) = +\infty$ and $h : X \to Z^\bullet = Z \cup \{\infty_C\}$ is a proper and C-convex function such that $h(\operatorname{dom} f \cap \operatorname{dom} h) \cap \operatorname{dom} g \neq \emptyset$. We use two perturbation functions in order to formulate corresponding conjugate duals to (P^{CC}). By particularizing the results from Section 1, we also give generalized interior point regularity conditions and strong duality theorems for both primal–dual pairs. One of the dual problems to (P^{CC}) that we obtain in this way is nothing else than the one considered by Combari, Laghdir and Thibault in [57]. Some of the regularity conditions we provide in connection with this duality concept slightly generalizes some conditions introduced in [57], while for the origins of the other ones we refer to the works of Zălinescu (cf. [125, 127]).

Chapter II. The main goal of the second chapter is to furnish generalizations of the celebrated Moreau–Rockafellar theorem from which so-called *closedness-type regularity conditions* for the problems treated in Chapter I are derived. In Section 5 we show that in case $\Phi : X \times Y \to \overline{\mathbb{R}}$ is a proper, convex and lower semicontinuous function such that $0 \in \operatorname{Pr}_Y(\operatorname{dom} \Phi)$, one has for all $x^* \in X^*$

$$(\Phi(\cdot, 0))^*(x^*) = \operatorname{cl}_{\omega*}\left(\inf_{y^* \in Y^*} \Phi^*(\cdot, y^*)\right)(x^*),$$

the lower semicontinuous hull being taken in the weak* topology $\omega(X^*, X)$ on X^*. This formula, which is a consequence of a result from Precupanu's paper [110], is said to be a *generalized Moreau–Rockafellar formula*. As a consequence of it we obtain the following characterization expressed by means of epigraphs

$$\operatorname{epi}((\Phi(\cdot, 0))^*) = \operatorname{cl}_{\omega^* \times \mathcal{R}}\left(\operatorname{epi}\left(\inf_{y^* \in Y^*} \Phi^*(\cdot, y^*)\right)\right) = \operatorname{cl}_{\omega^* \times \mathcal{R}}\left(\operatorname{Pr}_{X^* \times \mathbb{R}}(\operatorname{epi} \Phi^*)\right),$$

the closure being taken in $(X^*, \omega(X^*, X)) \times \mathbb{R}$. This means that the assumption $\operatorname{Pr}_{X^* \times \mathbb{R}}(\operatorname{epi} \Phi^*)$ is closed in $(X^*, \omega(X^*, X)) \times \mathbb{R}$ is a necessary and sufficient condition, called *closedness-type regularity condition*, for having

$$(\Phi(\cdot, 0))^*(x^*) = \min_{y^* \in Y^*} \Phi^*(x^*, y^*) \ \forall x^* \in X^*.$$

This situation is known in the literature under the name *stable strong duality*. Consequently, the above closedness-type condition turns out to be sufficient for having strong duality for (PG) and (DG). In Theorem 5.5 we state that the generalized interior point regularity conditions formulated in Section 1 for the primal–dual pair (PG)–(DG) are sufficient for stable strong duality, too, and, at the same time, that they imply the mentioned closedness-type regularity condition. Closedness-type conditions have been first introduced for the optimization problem minimizing the sum of two functions and its Fenchel dual problem by Burachik and Jeyakumar in [46–48]. For the optimization problem with geometric and cone constraints and its

Lagrange dual closedness-type conditions have been first considered by Jeyakumar, Dinh and Lee in [84] and Jeyakumar, Song, Dinh and Lee in [86]. These considerations are made for problems stated in Banach spaces and partially under quite strong topological assumptions for the functions involved. Under the pioneering works on closedness-type regularity conditions for different classes of optimization problems and corresponding conjugate duals, we mention here also the papers of Boţ and Wanka [39, 40] and Boţ, Grad and Wanka [24, 29]. In this book, we work in a more general setting and show how the closedness-type conditions from the mentioned literature relate to the ones we derive from the general ones when dealing with the corresponding particular instances of the general optimization problem (PG). But first, we provide in Section 6 by means of the perturbation functions considered in Section 4 in the context of the composed convex optimization problem (P^{CC}) generalized Moreau–Rockafellar results and from here closedness-type regularity conditions for both dual problems introduced there. To this aim, we consider the supplementary assumptions that f and g are lower semicontinuous and h is star C-lower semicontinuous. Example 6.5 presents a situation where the generalized interior point conditions are not fulfilled unlike the closedness-type ones. These achievements are based on investigations due to Boţ, Grad and Wanka from [27]. By employing Theorems 6.6 and 6.11, one can deduce that the closedness-type regularity conditions turn out to be necessary and sufficient for corresponding *ε-subdifferential sum formulae*, fact which was pointed out in [33]. Similar assertions can be obtained by using the perturbation functions considered in Sections 2 and 3, but we derive them in the forthcoming sections as particular instances of the composed convex case. More precisely, in Section 7 we deliver Moreau–Rockafellar formulae for $f + g \circ A$, where f, g and A are considered as in the definition of (P^A), this time by taking in the composed convex setting $Z = Y$, $C = \{0\}$ and $h : X \to Y$, $h(x) = Ax, x \in X$. In this way we rediscover some results of Boţ and Wanka published in [40]. We further consider three particular instances, one of them when $X = Y$ and A is the identity on X. In this case we obtain the classical *Moreau–Rockafellar Formula* $(f+g)^* = \mathrm{cl}_{\omega^*}(f^* \Box g^*)$ while the closedness-type condition becomes: $\operatorname{epi} f^* + \operatorname{epi} g^*$ is closed in $(X^*, \omega(X^*, X)) \times \mathbb{R}$. The latter is proved to be necessary and sufficient for having

$$(f+g)^*(x^*) = \min_{y^* \in Y^*} \{f^*(x^* - y^*) + g^*(y^*)\} \; \forall x^* \in X^*$$

or, equivalently,

$$\partial_\varepsilon (f+g)(x) = \bigcup_{\substack{\varepsilon_1, \varepsilon_2 \geq 0 \\ \varepsilon_1 + \varepsilon_2 = \varepsilon}} (\partial_{\varepsilon_1} f(x) + \partial_{\varepsilon_2} g(x)) \; \forall \varepsilon \geq 0 \; \forall x \in X.$$

This equivalence was first pointed out by Burachik and Jeyakumar in [47] in Banach spaces and follows as a particular instance of the general scheme. In Section 8, we turn our attention to the problem with geometric and cone constraints and provide generalized Moreau–Rockafellar formulae for the conjugate of $f + \delta_{\{y \in S: g(y) \in -C\}}$ along with corresponding regularity conditions of closedness-type for (P^C) and its Lagrange, Fenchel and Fenchel–Lagrange dual problems. By these particularizations, the statements given in Section 6 become the ones from [24] concerning regularity conditions and the ones from [33] concerning their connections to subdifferential formulae, respectively. We also study the relations among the considered closedness-type conditions and show by some examples that they are weaker than the corresponding generalized interior point ones. The last section of the chapter, Section 9, introduces the notion of closedness regarding a set. For U and V two subsets of a separated locally convex space X we say that U *is closed regarding the set* V if $U \cap V = U \cap \text{cl}(V)$. Under the hypotheses assumed for Φ in Section 5, when $U \subseteq X^*$ is a nonempty subset one has

$$(\Phi(\cdot, 0))^*(x^*) = \min_{y^* \in Y^*} \Phi^*(x^*, y^*) \ \forall x^* \in U$$

if and only if $\text{Pr}_{X^* \times \mathbb{R}}(\text{epi}\, \Phi^*)$ is closed regarding the set $U \times \mathbb{R}$ in $(X^*, \omega(X^*, X)) \times \mathbb{R}$. Introducing such a notion turns out to have a positive effect on the characterization of stable strong duality, especially in the case when the perturbation function Φ fails to be lower semicontinuous. As a consequence, one can provide closedness-type regularity conditions for (P^{CC}) and its two duals also in case when h is (only) C-epi closed, as it has been noticed by Boţ, Grad and Wanka in [29]. The statements we give in this setting substantially improve the ones stated by Combari, Laghdir and Thibault in [58]. On the other hand, we provide a necessary and sufficient condition for having an exact formula for the conjugate of the infimal value function of a convex function through a *convex process*. We notice that generalized interior point conditions guaranteeing this formula can be found in [127, Theorem 2.8.6].

Chapter III. Working in real normed spaces, we deal in this chapter with formulae for the biconjugates of different convex functions on X. In Theorem 10.3, the main result of Section 10, we show that if X and Y are normed spaces and $\Phi : X \times Y \to \overline{\mathbb{R}}$ is a proper, convex and lower semicontinuous function such that $0 \in \text{Pr}_Y(\text{dom}\, \Phi)$, then $(\Phi(\cdot, 0))^{**} = \Phi^{**}(\cdot, 0)$ if and only if the lower semicontinuous hulls of $\inf_{y^* \in Y^*} \Phi^*(\cdot, y^*)$ in the weak* and strong topologies of X^* coincide. In Theorem 10.4 sufficient conditions, which guarantee this equality, are given (see also [21]). In the next section, these achievements are applied to the different classes of convex functions considered in the previous chapters and corresponding formulae for their biconjugates are given. In this way, we rediscover and partially extend some results given by Zălinescu in [129]. In Section 12 the supremum of an (infinite) family of proper, convex and lower semicontinuous functions $f_t : X \to \overline{\mathbb{R}}, t \in T$, is considered. For the conjugate of this supremum Moreau–Rockafellar-type formulae along with different characterizations of its epigraph are provided. We extend the investigations to the biconjugate of the supremum and close the section with Theorem 12.6, where we formulate sufficient weak conditions for having $(\sup_{t \in T} f_t)^{**} = \sup_{t \in T} f_t^{**}$. The case $T = 2$ is treated separately in Section 13. There we give also a refined formula for the conjugate of the supremum of two functions and show that the equality above is true under the classical Attouch–Brézis regularity condition. With this result, we propose an alternative proof of a statement which is due to Fitzpatrick and Simons (see [71]).

Chapter IV. In this chapter, we weaken the closedness-type regularity conditions and obtain a further class of conditions which does not necessarily ensure stable strong duality, but are still sufficient for strong duality. We make this in Section 14 for the primal–dual pair (PG)–(DG), as well as for the composed convex optimization problem (P^{CC}) and its conjugate duals formulated in Section 4, by assuming for the latter that h is star C-lower semicontinuous. For similar investigations made in case h is C-epi closed we refer the reader to [29]. In the next section we particularize these regularity conditions to the case of the primal–dual pair (P^A)–(D^A) and to further special instances of it. Example 15.3, which has been first considered in [40], presents a situation where this new condition is fulfilled unlike the closedness-type one stated in Section 7. A converse duality theorem for (P^A)–(D^A) is also formulated and proved. The approach we consider is based on a fruitful idea used by Bauschke in [4] and later by Ng and Song in [102]. Regularity conditions, weaker than the closedness-type ones, for (P^C) and its Lagrange and Fenchel–Lagrange duals are considered in Section 16 (see also [24, 39]). By a suitable example, we show that there exist situations where the first ones are fulfilled while the latter fail. A comparison of the new regularity conditions with similar sufficient conditions for Lagrange duality stated by Jeyakumar and its co-authors in [84, 86] is also made. As a particular case we rediscover the so-called *Farkas–Minkowski condition* for a system of (infinitely many) convex inequalities, which comes from the literature dealing with semi-infinite programming problems (see, for instance, [64, 74, 75]). In the lines of the investigations made by Boţ, Grad and Wanka in [30, 31], in Section 17 we continue treating the primal problem (P^C) and its Lagrange and Fenchel–Lagrange duals and consider the situation called *total duality*. This is the situation when strong duality holds and the primal problem has an optimal solution, too. We characterize it via some formulae expressed by using the subdifferentials of the functions involved, which we call *generalized Basic Constraint Qualifications (gBCQ)*, as they can be seen as extensions of the classical *Basic Constraint Qualifications* (BCQ). For the original formulation of the latter we refer to the book of Hiriart-Urruty and Lemaréchal [79]. When the cone constraints in the formulation of the primal problem are replaced by (infinitely many) inequality constraints $(gBCQ)$ becomes the so-called *locally Farkas–Minkowski condition*, which also occurs in a considerable number of investigations on infinite and semi-infinite programming problems (see for instance [64, 65, 69, 74]).

Chapter V. The fifth chapter of this work, where we offer some deeper insights into Fenchel duality, is divided into two parts. In the first part we deal with so-called *totally Fenchel unstable functions*, while in the second part regularity conditions expressed via the *quasi interior* and *quasi-relative interior* are addressed. If X is a separated locally convex space and $f, g : X \to \overline{\mathbb{R}}$ are two arbitrary proper functions, we say that the pair f, g is *totally Fenchel unstable* if f and g satisfy Fenchel duality but

$$y^*, z^* \in X^* \text{ and } (f+g)^*(y^*+z^*) = f^*(y^*) + g^*(z^*) \text{ implies } y^* + z^* = 0.$$

This notion has been recently introduced by Stephen Simons in [120]. In Section 18 we deliver some characterizations of this situation and discuss an example of such a pair of functions given in [120]. Corollary 18.9 gives an answer to Problem 11.5 posed by Simons in [120] in the context of this example. To this end we appeal to the notion of a weak*-extreme point of a set, which appears in different results having their origins in the geometric functional analysis. An alternative solution for this problem by using a minmax theorem is also provided and is based on the investigations done by Boţ and Csetnek in [16]. Further, in Section 19, we give a negative answer to Problem 11.6 in [120] and prove actually that there is no pair of proper and convex totally Fenchel unstable functions if X is a nonzero finite dimensional space. In the second part of Chapter V, our goal is to deliver generalized interior point regularity conditions for both Fenchel and Lagrange duality formulated via the quasi interior and quasi-relative interior, the latter being considered first by Borwein and Lewis in [11]. In Section 20 these two notions are introduced and characterized. Separation results for convex sets under some conditions expressed via these two generalized interior notions are proved. They extend some similar statements proven in [59, 60]. An important role in the following investigations is played by a separation result of the same type stated by Cammaroto and Di Bella in [54]. Theorem 21.5, which is the main result of Section 21, furnishes the announced sufficient conditions for Fenchel duality. This theorem is followed by other theorems and corollaries which fulfill the same task. In Example 21.10 we present a problem for which the classical generalized interior point conditions are not verified, but the strong duality theorems given in this section are applicable. These investigations are based on the work of Boţ, Csetnek and Wanka in [20]. In the next section, we extend the investigations to the optimization problem with geometric and cone constraints (P^C) and its Lagrange dual. Corresponding regularity conditions and strong duality theorems are derived from the ones obtained in Section 20 by using an approach due to Magnanti (cf. [95]). The strong duality theorems stated here improve and correct some results recently given in the literature (see, for instance, [54, 59, 60]). For more details on this topic, we refer to the investigations due to Boţ, Csetnek and Moldovan in [17].

Chapter VI. The last chapter is dedicated to the applications of the conjugate duality in the field of monotone operators. Some preliminary notions and results are introduced in Section 23 along with the notions of *Fitzpatrick function* and *representative function* associated to a monotone operator. Through these two objects, which have been considered first by Fitzpatrick (see [70]), one can connect the theory of monotone operators to results coming from the convex analysis. Neglected for many years until re-popularized in [7, 8, 52, 98, 104–106, 121], they have given rise to a great number of publications which rediscovered and extended the important results of the theory of monotone operators by using tools from the convex analysis. In the next section we consider X and Y reflexive Banach spaces, $S : X \rightrightarrows X^*$, $T : Y \rightrightarrows Y^*$ maximal monotone operators with representative functions $h_S : X \times X^* \to \overline{\mathbb{R}}$ and $h_T : Y \times Y^* \to \overline{\mathbb{R}}$, respectively, and $A : X \to Y$

a linear continuous operator such that $A(\Pr_X(\operatorname{dom} h_S)) \cap \Pr_X(\operatorname{dom} h_T) \neq \emptyset$. We give a closedness-type regularity condition, called (RC^{S+A^*TA}), (see also [18,23]) expressed by means of the conjugate functions of h_S and h_T which ensures that $S + A^*TA$ is a maximal monotone operator. We prove that (RC^{S+A^*TA}) is implied by the generalized interior point regularity conditions stated in the literature which fulfill this task, too, as are the ones given by Pennanen in [103] and Penot and Zălinescu in [108]. The maximal monotonicity of A^*TA and, in case $X = Y$ and A is the identity on X, of $S + T$ is considered as a particular case in Section 25. In both situations, we give examples of maximal monotone operators for which the generalized interior point conditions, as are in the second particular instance the ones due to Rockafellar (see [115]), Borwein (see [7]), Penot (see [105]), Chu (see [56]), Pennanen (see [103]), Penot and Zălinescu (see [108, 128]), fail while the closedness-type conditions are verified. Two open problems regarding this topic are formulated. In Section 26, we deal with *enlargements* of monotone operators defined by means of corresponding representative functions in general Banach spaces, a notion inspired by the ε-subdifferential of a proper, convex and lower semicontinuous function and which is due to Burachik and Svaiter (see [51,53]). More precisely, in the same lines of the investigations made by Boţ and Csetnek in [15] we give a so-called *bivariate infimal convolution formula* which turns out to be necessary and sufficient for having a sum formula for enlargements of maximal monotone operators. As a special case, we rediscover the equivalence between the condition that $\operatorname{epi} f^* + \operatorname{epi} g^*$ is closed in $(X^*, \omega(X^*, X)) \times \mathbb{R}$ and the fulfillment of the ε-subdifferential sum formula obtained in Section 7 when $f, g : X \to \overline{\mathbb{R}}$ are proper, convex and lower semicontinuous functions with $\operatorname{dom} f \cap \operatorname{dom} g \neq \emptyset$.

Chapter I
Perturbation Functions and Dual Problems

1 A General Approach for Duality

The starting point of our investigations is a general approach for constructing a dual optimization problem to a primal one based on the theory of conjugate functions. Consider X a separated locally convex spaceand $F : X \to \overline{\mathbb{R}} = \mathbb{R} \cup \{\pm\infty\}$ a given function. We assume that F is *proper*, namely that $F(x) > -\infty$ for all $x \in X$ and its *domain* $\operatorname{dom} F = \{x \in X : F(x) < +\infty\}$ is nonempty. To the general optimization problem, called also *primal problem*,

$$(PG) \quad \inf_{x \in X} F(x)$$

one can assign a *conjugate dual problem* by making use of the so-called *perturbation approach*. To this end, we consider Y another separated locally convex space and the function $\Phi : X \times Y \to \overline{\mathbb{R}}$, called *perturbation function*, fulfilling $\Phi(x, 0) = F(x)$ for all $x \in X$. The initial problem (PG) is nothing else than

$$(PG) \quad \inf_{x \in X} \Phi(x, 0).$$

Let X^* and Y^* be the topological dual spaces of the space of the *primal variables* X and the space of the *perturbation variables* Y, respectively. Assume in the following that both dual spaces are endowed with the weak* topology, denoted by $\omega(X^*, X)$ and $\omega(Y^*, Y)$, respectively. We denote by $\langle \cdot, \cdot \rangle$ the duality product in $X^* \times X$, i.e. for $x^* \in X$ and $x \in X$ we have $\langle x^*, x \rangle := x^*(x)$. Having a function $f : X \to \overline{\mathbb{R}}$, $f^* : X^* \to \overline{\mathbb{R}}$ defined by $f^*(x^*) = \sup_{x \in X}\{\langle x^*, x \rangle - f(x)\}$ will be the *conjugate function* of f. For the function f and its conjugate f^* the so-called *Young–Fenchel inequality* is always fulfilled: $f^*(x^*) + f(x) \geq \langle x^*, x \rangle$ for all $x \in X$ and $x^* \in X^*$.

A *conjugate dual problem* to (PG) can be formulated as being

$$(DG) \quad \sup_{y^* \in Y^*} \{-\Phi^*(0, y^*)\},$$

R.I. Boţ, *Conjugate Duality in Convex Optimization*, Lecture Notes in Economics and Mathematical Systems 637, DOI 10.1007/978-3-642-04900-2_1,

where $\Phi^* : X^* \times Y^* \to \overline{\mathbb{R}}$ is the conjugate function of Φ. Further, we denote by $v(PG)$ and $v(DG)$ the optimal objective values of the problems (PG) and (DG), respectively. The next result shows that *weak duality*, namely the fact that the optimal objective value of the primal problem is always greater than or equal to the optimal objective value of the dual problem, is a consequence of the way in which the latter problem is defined.

Theorem 1.1. *It holds*

$$-\infty \leq v(DG) \leq v(PG) \leq +\infty.$$

Proof. For all $x \in X$ and all $y^* \in Y^*$, by the Young–Fenchel inequality, we have

$$\Phi(x,0) + \Phi^*(0,y^*) \geq \langle 0,x \rangle + \langle y^*,0 \rangle = 0 \Leftrightarrow \Phi(x,0) \geq -\Phi^*(0,y^*),$$

which implies that $v(PG) \geq v(DG)$. □

Next, we characterize the existence of *strong duality*, namely of the situation when $v(PG) = v(DG)$ and the dual has an optimal solution. To this end, we consider the *infimal value function* of Φ, $h : Y \to \overline{\mathbb{R}}$ defined by $h(y) = \inf\{\Phi(x,y) : x \in X\}$. One can notice that $v(PG) = h(0)$. As follows from the next result, the optimal objective value of (DG) can be expressed by using the biconjugate of h. This is defined as $h^{**} = (h^*)^*$.

Proposition 1.2. *It holds* $v(DG) = h^{**}(0)$.

Proof. For all $y^* \in Y^*$ there is

$$h^*(y^*) = \sup_{y \in Y}\{\langle y^*,y \rangle - h(y)\} = \sup_{\substack{x \in X \\ y \in Y}}\{\langle y^*,y \rangle - \Phi(x,y)\} = \Phi^*(0,y^*). \quad (1.1)$$

Thus

$$h^{**}(0) = \sup_{y^* \in Y^*}\{-h^*(y^*)\} = \sup_{y^* \in Y^*}\{-\Phi^*(0,y^*)\} = v(DG).$$

□

Since $h^{**}(x) \leq h(x)$ for all $x \in X$, one has that $h^{**}(0) \leq h(0)$, which is nothing else than the relation that states the existence of weak duality, i.e. $v(DG) \leq v(PG)$.

Considering a function $f : X \to \overline{\mathbb{R}}$, we say that f is *convex* if for all $x, y \in X$ and all $\lambda \in [0,1]$ it holds $f(\lambda x + (1-\lambda)y) \leq \lambda f(x) + (1-\lambda)f(y)$, whenever the following conventions are considered: $(+\infty) + (-\infty) = (-\infty) + (+\infty) = +\infty$, $0(+\infty) = +\infty$ and $0(-\infty) = 0$. The *epigraph* of the function f is the set $\operatorname{epi} f = \{(x,r) \in X \times \mathbb{R} : f(x) \leq r\}$ and we have that f is convex if and only if its epigraph is a convex set. In this situation $\operatorname{dom} f$ is a convex set, too. Having $U \subseteq X$ a given set, we denote by $\operatorname{int}(U)$ and $\operatorname{cl}(U)$ the topological *interior* and *closure* of the set U, respectively. The *indicator function* of U is $\delta_U : X \to \overline{\mathbb{R}}$,

$$\delta_U(x) = \begin{cases} 0, & \text{if } x \in U, \\ +\infty, & \text{otherwise.} \end{cases}$$

Its conjugate function is the *support function* of U, $\sigma_U : X^* \to \overline{\mathbb{R}}$, $\sigma_U(x^*) = \sup_{u \in U} \langle x^*, u \rangle$. Further, we denote by $f_U^* : X^* \to \overline{\mathbb{R}}$, $f_U^*(x^*) = \sup_{x \in U}\{\langle x^*, x\rangle - f(x)\} = (f + \delta_U)^*(x^*)$ the *conjugate of* f *with respect to the set* U. According to the conventions made above, we treat $0f$ as being equal to $\delta_{\operatorname{dom} f}$. The *lower semicontinuous hull* of the function f is the function $\operatorname{cl} f : X \to \overline{\mathbb{R}}$, the epigraph of which is equal to $\operatorname{cl}(\operatorname{epi} f)$. By the *Fenchel–Moreau Theorem* (see, for instance, [127, Theorem 2.3.4]) one has that whenever $f : X \to \overline{\mathbb{R}}$ is a convex function such that $\operatorname{cl} f$ is proper, then $f^{**} = \operatorname{cl} f$. We also want to notice that $\operatorname{cl} f$ is proper if and only if f^* is proper.

Definition 1.3. We say that the problem (PG) is *normal* if $h(0) \in \mathbb{R}$ and h is lower semicontinuous at 0.

The next result provides a characterization of the normality of (PG).

Theorem 1.4 (cf. [67]). *Assume that* $\Phi : X \times Y \to \overline{\mathbb{R}}$ *is a proper and convex function. Then the following statements are equivalent:*

(i) The problem (PG) *is normal.*
(ii) It holds $v(PG) = v(DG)$ *and this value is finite.*

Proof. (i) $\Rightarrow$ (ii) We notice first that

$$h^{**}(y) \le \operatorname{cl} h(y) \le h(y) \ \forall y \in Y. \tag{1.2}$$

Since Φ is convex, h is also convex and this implies that $\operatorname{cl} h$ is convex, too. The problem (PG) being normal, it follows that $\operatorname{cl} h(0) = h(0) \in \mathbb{R}$. Since $\operatorname{cl} h$ is a convex and lower semicontinuous function, we have that $\operatorname{cl} h(y) > -\infty$ for all $y \in Y$ (see, for instance, [127, Proposition 2.2.5]). This guarantees the properness of $\operatorname{cl} h$. Taking into consideration that the conjugate of a function is equal to the conjugate of its lower semicontinuous hull, by the Fenchel–Moreau Theorem, we obtain $\operatorname{cl} h = (\operatorname{cl} h)^{**} = ((\operatorname{cl} h)^*)^* = (h^*)^* = h^{**}$ and so $h^{**}(0) = \operatorname{cl} h(0) = h(0) \in \mathbb{R}$. Since $v(PG) = h(0)$ and $v(DG) = h^{**}(0)$, (ii) is true.

(ii) $\Rightarrow$ (i) The statement (ii) can be equivalently written as $h^{**}(0) = h(0) \in \mathbb{R}$. Then, by (1.2), we get that $\operatorname{cl} h(0) = h(0) \in \mathbb{R}$, which means that (PG) is normal. □

In order to be able to characterize the strong duality for (PG) and (DG) we need to introduce some further elements from convex analysis. Consider $f : X \to \overline{\mathbb{R}}$ a given function and $\varepsilon \ge 0$. If $x \in X$ is such that $f(x) \in \mathbb{R}$, the *ε-subdifferential* of f at x is the set $\partial_\varepsilon f(x) = \{x^* \in X^* : f(y) - f(x) \ge \langle x^*, y - x\rangle - \varepsilon \ \forall y \in X\}$. Otherwise, we set $\partial_\varepsilon f(x) := \emptyset$. We denote by $\partial f(x) := \partial_0 f(x)$ the *subdifferential* of f at x. If $\partial f(x) \ne \emptyset$, then f is said to be *subdifferentiable* at x. The ε-subdifferential of the indicator function of a nonempty set $U \subseteq X$ at $x \in U$,

$N_U^\varepsilon(x) := \partial_\varepsilon(\delta_U)(x) = \{x^* \in X^* : \langle x^*, y - x\rangle \leq \varepsilon \ \forall y \in U\}$, is called the *ε-normal set* of the set U at x. The *normal cone* of the set U at $x \in U$ is $N_U(x) := N_U^0(x)$.

The elements $x^* \in \partial f(x)$ are called *subgradients* of f. One can easily verify that for all $\varepsilon \geq 0$

$$x^* \in \partial_\varepsilon f(x) \Leftrightarrow f^*(x^*) + f(x) \leq \langle x^*, x\rangle + \varepsilon, \tag{1.3}$$

which has as consequence that

$$x^* \in \partial f(x) \Leftrightarrow f^*(x^*) + f(x) = \langle x^*, x\rangle. \tag{1.4}$$

By means of the ε-subdifferential one can characterize the ε-optimal solutions of the problem (PG). For $\varepsilon \geq 0$ we call $\bar{x} \in \operatorname{dom} F$ an *ε-optimal solution* for (PG) if $F(\bar{x}) - \varepsilon \leq F(x)$ for all $x \in X$, which is the same with saying that $0 \in \partial_\varepsilon F(\bar{x})$. For $\varepsilon = 0$ we get that $\bar{x} \in \operatorname{dom} F$ is an *optimal solution* for (PG) if and only if $0 \in \partial F(\bar{x})$.

Definition 1.5. We say that the problem (PG) is *stable* if $h(0) \in \mathbb{R}$ and h is subdifferentiable at 0.

Theorem 1.6 (cf. [67]). *Assume that $\Phi : X \times Y \to \overline{\mathbb{R}}$ is a proper and convex function. Then the following statements are equivalent:*

(i) The problem (PG) is stable.
(ii) The primal problem (PG) is normal and the dual (DG) has an optimal solution. In this situation the set of optimal solutions of (DG) is equal to $\partial h(0)$.

Proof. (i) $\Rightarrow$ (ii) Assume that $h(0) \in \mathbb{R}$ and $\partial h(0) \neq \emptyset$. Thus $h(0) = h^{**}(0)$ (cf. [127, Theorem 2.4.1]), which is nothing else than $v(PG) = v(DG) \in \mathbb{R}$. By Theorem 1.4 follows that (PG) is normal. Further, consider an element $\bar{y}^* \in \partial h(0)$. We have $h(0) + h^*(\bar{y}^*) = 0$ or, equivalently, $v(PG) = h(0) = -h^*(\bar{y}^*) = -\Phi^*(0, \bar{y}^*)$ and, by Theorem 1.1, it follows that $\bar{y}^*$ is an optimal solution for (DG).

(ii) $\Rightarrow$ (i) Assume that (PG) is normal and the dual (DG) has an optimal solution $\bar{y}^* \in Y^*$. Thus $h(0) = v(PG) = v(DG) = -\Phi^*(0, \bar{y}^*) = -h^*(\bar{y}^*) \in \mathbb{R}$, which is the same with $h(0) + h^*(\bar{y}^*) = 0 \Leftrightarrow \bar{y}^* \in \partial h(0)$. Since the set $\partial h(0)$ is nonempty, the stability of (PG) is guaranteed. □

As we have seen above, the stability completely characterizes the existence of strong duality for a given convex optimization problem and its conjugate dual. Therefore, one of the main issues in the optimization theory is to formulate sufficient conditions, called *regularity conditions*, which ensure that the primal problem is stable. We formulate in the following some so-called *generalized interior point regularity conditions* by using the perturbation function Φ which we considered in this section. To this end we need to introduce some extensions of the classical interior of a set.

Having U a subset of X, we denote by $\text{lin}(U)$, $\text{aff}(U)$, $\text{co}(U)$, $\text{cone}(U)$ and $\text{coneco}(U)$ its *linear hull*, *affine hull*, *convex hull*, *conic hull* and *convex conic hull*, respectively. The *algebraic interior* of U is the set $\text{core}(U) = \{u \in U : \forall x \in X, \exists \delta > 0 \text{ such that } \forall \lambda \in [0, \delta] : u + \lambda x \in U\}$, while the *intrinsic core* of U is the set (cf. [77]) $\text{icr}(U) = \{u \in U : \forall x \in \text{aff}(U - U), \exists \delta > 0 \text{ such that } \forall \lambda \in [0, \delta] : u + \lambda x \in U\}$. When $U \subseteq X$ is a convex set, then $u \in \text{core}(U) \Leftrightarrow \text{cone}(U - u) = X$, while $u \in \text{icr}(U) \Leftrightarrow \text{cone}(U - u)$ is a linear subspace. One has that $\text{int}(U) \subseteq \text{core}(U) \subseteq \text{icr}(U)$. In case U is convex and one of the following conditions is fulfilled: $\text{int}(U) \neq \emptyset$; X is Banach and U is closed; X is finite dimensional, then $\text{int}(U) = \text{core}(U)$ (cf. [116]). For a convex set with empty interior and nonempty algebraic interior we refer to [13, Exercise 4.1.9]. Further, by maintaining the convexity assumption for U one can consider the *strong quasi-relative interior* of this set, denoted by $\text{sqri}(U)$, as being the set of those $u \in U$ for which $\text{cone}(U - u)$ is a closed linear subspace (cf. [10, 11]). We have that $\text{core}(U) \subseteq \text{sqri}(U) \subseteq \text{icr}(U)$ and due to [76] it holds $u \in \text{sqri}(U)$ if and only if $u \in \text{icr}(U)$ and $\text{aff}(U - u)$ is a closed linear subspace. Assuming additionally that X is finite dimensional, it yields $\text{icr}(U) = \text{sqri}(U) = \text{ri}(U)$, where by $\text{ri}(U)$ we denote the *relative interior* of the set U. This is the set of the interior points of U relative to the affine hull of U.

Assume in the following that the perturbation function $\Phi : X \times Y \to \overline{\mathbb{R}}$ is proper and convex such that $0 \in \text{Pr}_Y(\text{dom}\,\Phi)$. Here $\text{Pr}_Y : X \times Y \to Y$, defined by $\text{Pr}_Y(x, y) = y, (x, y) \in X \times Y$, is the *projection operator* on Y. The first regularity condition which we state here is formulated in the general setting where X and Y are separated locally convex spaces (cf. [67]):

(RC_1^Φ) | $\exists x' \in X$ such that $(x', 0) \in \text{dom}\,\Phi$ and $\Phi(x', \cdot)$ is continuous at 0.

For the forthcoming regularity conditions we assume that X and Y are Fréchet spaces (cf. [116, 126, 127]):

(RC_2^Φ) | X and Y are Fréchet spaces, Φ is lower semicontinuous and $0 \in \text{sqri}(\text{Pr}_Y(\text{dom}\,\Phi))$.

The regularity condition (RC_2^Φ) has been introduced as the weakest one in a succession of regularity conditions involving different generalized interior notions (cf. [116]):

$(RC_{2'}^\Phi)$ | X and Y are Fréchet spaces, Φ is lower semicontinuous and $0 \in \text{core}(\text{Pr}_Y(\text{dom}\,\Phi))$,

respectively,

$(RC_{2''}^\Phi)$ | X and Y are Fréchet spaces, Φ is lower semicontinuous and $0 \in \text{int}(\text{Pr}_Y(\text{dom}\,\Phi))$.

Regarding the last two conditions, we want to make the following comment. In case Φ is convex and lower semicontinuous its infimal value function $h : Y \to \overline{\mathbb{R}}$, $h(y) = \inf_{x \in X} \Phi(x, y)$, is convex but not necessarily lower semicontinuous, fulfilling $\text{dom}\,h = \text{Pr}_Y(\text{dom}\,\Phi)$. Nevertheless, when X and Y are Fréchet spaces, by [127, Proposition 2.2.18] the function h is *li-convex*. Using [127, Theorem 2.2.20] it follows that core(dom h) = int(dom h), which has as consequence the equivalence of the regularity conditions $(RC_{2'}^{\Phi})$ and $(RC_{2''}^{\Phi})$. Thus, $(RC_{2''}^{\Phi}) \Leftrightarrow (RC_{2'}^{\Phi}) \Rightarrow (RC_{2}^{\Phi})$, all these conditions being implied by (RC_{1}^{Φ}) when X and Y are Fréchet spaces and Φ is lower semicontinuous. One should notice that the relation $0 \in \operatorname{sqri}(\Pr_Y(\operatorname{dom}\Phi))$ in (RC_2^{Φ}) can be replaced by the equivalent formulation $0 \in \operatorname{icr}(\Pr_Y(\operatorname{dom}\Phi))$ and $\operatorname{aff}(\Pr_Y(\operatorname{dom}\Phi))$ is a closed linear subspace (cf. [76, Proposition 3.4]). The following generalized interior point regularity condition is of interest especially when dealing with convex optimization problems in finite dimensional spaces (cf. [113, 127]):

$$(RC_3^{\Phi}) \mid \dim(\operatorname{lin}(\Pr_Y(\operatorname{dom}\Phi))) < +\infty \text{ and } 0 \in \operatorname{ri}(\Pr_Y(\operatorname{dom}\Phi)).$$

By collecting the corresponding results from [67, 113, 116, 127] one can state the following strong duality theorem for the primal–dual pair $(PG) - (DG)$.

Theorem 1.7. *Let $\Phi : X \times Y \to \overline{\mathbb{R}}$ be a proper and convex function such that $0 \in \Pr_Y(\operatorname{dom}\Phi)$. If one of the regularity conditions (RC_i^{Φ}), $i \in \{1, 2, 3\}$, is fulfilled, then $v(PG) = v(DG)$ and the dual has an optimal solution.*

Remark 1.8. One should notice that in the result above if one of the regularity conditions (RC_i^{Φ}), $i \in \{1, 2, 2', 2'', 3\}$, is satisfied, then the feasibility assumption $0 \in \Pr_Y(\operatorname{dom}\Phi)$ is automatically fulfilled. The same applies for the strong duality theorems that we state in the forthcoming sections for different particular instances of the general primal–dual pair (PG)–(DG). Nevertheless, we opted for imposing the feasibility assumption when formulating the primal problem in order to avoid dealing with degenerate optimization problems.

2 The Problem Having the Composition with a Linear Continuous Operator in the Objective Function

By using the general approach described in the previous section we develop a conjugate duality theory and provide corresponding generalized interior point regularity conditions for the following optimization problem

$$(P^A) \quad \inf_{x \in X} \{f(x) + (g \circ A)(x)\}$$

and some particular instances of it. Here X and Y are separated locally convex spaces, $A : X \to Y$ is a linear continuous operator and $f : X \to \overline{\mathbb{R}}$ and $g : Y \to \overline{\mathbb{R}}$ are proper functions fulfilling $A(\operatorname{dom} f) \cap \operatorname{dom} g \neq \emptyset$. Consider as perturbation function to (P^A) $\Phi^A : X \times Y \to \overline{\mathbb{R}}$, $\Phi^A(x, y) = f(x) + g(Ax + y)$, where $y \in Y$ is the perturbation variable. Denote by $A^* : Y^* \to X^*$ the *adjoint operator* of A defined by $\langle A^* y^*, x\rangle = \langle y^*, Ax\rangle$ for all $(x, y^*) \in X \times Y^*$. The conjugate function of Φ^A, $(\Phi^A)^* : X^* \times Y^* \to \overline{\mathbb{R}}$, has for all $(x^*, y^*) \in X^* \times Y^*$ the following formulation

$$(\Phi^A)^*(x^*, y^*) = \sup_{\substack{x\in X\\ y\in Y}}\{\langle x^*, x\rangle + \langle y^*, y\rangle - f(x) - g(Ax + y)\} = \sup_{\substack{x\in X\\ r\in Y}}\{\langle x^*, x\rangle$$

$$+\langle y^*, r - Ax\rangle - f(x) - g(r)\} = \sup_{\substack{x\in X\\ r\in Y}}\{\langle x^* - A^*y^*, x\rangle + \langle y^*, r\rangle - f(x) - g(r)\}$$

$$= f^*(x^* - A^*y^*) + g^*(y^*). \tag{2.1}$$

The conjugate dual of (P^A) obtained by means of the perturbation function Φ^A looks like

$$(D^A) \quad \sup_{y^*\in Y^*}\{-f^*(-A^*y^*) - g^*(y^*)\}$$

and can be seen as a *Fenchel-type dual* problem to (P^A). By the weak duality theorem Theorem 1.1 follows $v(D^A) \le v(P^A)$.

Assuming additionally that f and g are convex functions, it follows that Φ^A is proper and convex, too, and the feasibility condition guarantees that $0 \in \Pr_Y(\operatorname{dom}\Phi^A) = \{y \in Y : \exists x \in \operatorname{dom} f \text{ such that } y \in \operatorname{dom} g - Ax\} = \operatorname{dom} g - A(\operatorname{dom} f)$. Thus the regularity condition (RC_1^Φ) can be equivalently written as

$$(RC_1^A) \mid \exists x' \in \operatorname{dom} f \cap A^{-1}(\operatorname{dom} g) \text{ such that } g \text{ is continuous at } Ax'.$$

When $V \subseteq Y$ is a given set, denote $A^{-1}(V) = \{x \in X : Ax \in V\}$. In Fréchet spaces one can deduce from the general case the following regularity conditions for the pair (P^A)–(D^A)

(RC_2^A) | X and Y are Fréchet spaces, f and g are lower semicontinuous and $0 \in \operatorname{sqri}(\operatorname{dom} g - A(\operatorname{dom} f))$

along with its stronger versions

$(RC_{2'}^A)$ | X and Y are Fréchet spaces, f and g are lower semicontinuous and $0 \in \operatorname{core}(\operatorname{dom} g - A(\operatorname{dom} f))$

and

$(RC_{2''}^A)$ | X and Y are Fréchet spaces, f and g are lower semicontinuous and $0 \in \operatorname{int}(\operatorname{dom} g - A(\operatorname{dom} f))$,

which are in fact equivalent.

Rockafellar considered $(RC_{2'}^A)$ in [112], while Rodrigues dealt with (RC_2^A) in [117] (see also the paper of Rodrigues and Simons [118]). Gowda and Teboulle have given in [76] an equivalent formulation for (RC_2^A) by means of the intrinsic core. In the finite dimensional setting one can use the following regularity condition

(RC_3^A) | $\dim(\operatorname{lin}(\operatorname{dom} g - A(\operatorname{dom} f))) < +\infty$ and $\operatorname{ri}(\operatorname{dom} g) \cap \operatorname{ri}(A(\operatorname{dom} f))) \neq \emptyset$,

which becomes in case $X = \mathbb{R}^n$ and $Y = \mathbb{R}^m$

$$(RC_{fin}^A) \mid \exists x' \in \operatorname{ri}(\operatorname{dom} f) \text{ such that } Ax' \in \operatorname{ri}(\operatorname{dom} g).$$

This is the classical regularity condition for this primal–dual pair in finite dimensional spaces and has been stated by Rockafellar in [113, Corollary 31.2.1].

We have the following strong duality theorem (see also [127]).

Theorem 2.1. *Let $f : X \to \overline{\mathbb{R}}$ and $g : Y \to \overline{\mathbb{R}}$ be proper and convex functions and $A : X \to Y$ be a linear continuous operator such that $A(\operatorname{dom} f) \cap \operatorname{dom} g \neq \emptyset$. If one of the regularity conditions (RC_i^A), $i \in \{1, 2, 3\}$, is fulfilled, then $v(P^A) = v(D^A)$ and the dual has an optimal solution.*

By taking $X = Y$ and $A = \mathrm{id}_X$, the identity operator on X, the primal problem becomes

$$(P^{\mathrm{id}}) \quad \inf_{x \in X} \{f(x) + g(x)\},$$

while the perturbation function $\Phi^{\mathrm{id}} : X \times X \to \overline{\mathbb{R}}$ is defined by $\Phi^{\mathrm{id}}(x, y) = f(x) + g(x + y)$. Its conjugate is given by the following formula

$$(\Phi^{\mathrm{id}})^*(x^*, y^*) = f^*(x^* - y^*) + g^*(y^*) \ \forall (x^*, y^*) \in X^* \times X^* \tag{2.2}$$

and the conjugate dual of (P^{id}) looks like

$$(D^{\mathrm{id}}) \quad \sup_{y^* \in X^*} \{-f^*(-y^*) - g^*(y^*)\},$$

which is actually the classical *Fenchel dual* problem to (P^{id}). The existence of weak duality for (P^{id}) and (D^{id}) follows from Theorem 1.1, too.

By particularizing the regularity conditions enunciated for the pair (P^A)–(D^A), they become

$$(RC_1^{\mathrm{id}}) \ \big| \ \exists x' \in \operatorname{dom} f \cap \operatorname{dom} g \text{ such that } g \text{ (or } f) \text{ is continuous at } x',$$

in case X is a Fréchet space

$$(RC_2^{\mathrm{id}}) \ \left| \begin{array}{l} X \text{ is a Fréchet space, } f \text{ and } g \text{ are lower semicontinuous} \\ \text{and } 0 \in \operatorname{sqri}(\operatorname{dom} g - \operatorname{dom} f) \end{array} \right.$$

along with its stronger versions

$$(RC_{2'}^{\mathrm{id}}) \ \left| \begin{array}{l} X \text{ is a Fréchet space, } f \text{ and } g \text{ are lower semicontinuous} \\ \text{and } 0 \in \operatorname{core}(\operatorname{dom} g - \operatorname{dom} f) \end{array} \right.$$

and

$$(RC_{2''}^{\mathrm{id}}) \ \left| \begin{array}{l} X \text{ is a Fréchet space, } f \text{ and } g \text{ are lower semicontinuous} \\ \text{and } 0 \in \operatorname{int}(\operatorname{dom} g - \operatorname{dom} f), \end{array} \right.$$

which are in fact equivalent.

The regularity condition (RC_2^{id}) was introduced by Attouch and Brézis in [2] in case X is a Banach space and bears their names. In the finite dimensional case one can consider

$$(RC_3^{\mathrm{id}}) \ \big| \ \dim(\operatorname{lin}(\operatorname{dom} g - \operatorname{dom} f)) < +\infty \text{ and } \operatorname{ri}(\operatorname{dom} g) \cap \operatorname{ri}(\operatorname{dom} f) \neq \emptyset,$$

which becomes when $X = \mathbb{R}^n$

$$(RC_{fin}^{\text{id}}) \;\Big|\; \text{ri}(\text{dom}\, f) \cap \text{ri}(\text{dom}\, g) \neq \emptyset,$$

in fact, the classical regularity condition considered by Rockafellar in [113, Theorem 31.1] (see also [12]) for proving the Fenchel Duality Theorem in finite dimensional spaces. One should notice that whenever f is polyhedral in the formulation of the regularity condition one can replace ri(dom f) by dom f. The same applies if g is polyhedral.

Theorem 2.2. *Let $f, g : X \to \overline{\mathbb{R}}$ be proper and convex functions such that* $\text{dom}\, f \cap \text{dom}\, g \neq \emptyset$. *If one of the regularity conditions (RC_i^{id}), $i \in \{1, 2, 3\}$, is fulfilled, then $v(P^{\text{id}}) = v(D^{\text{id}})$ and the dual has an optimal solution.*

We come now to a second special case of (P^A), namely the one when $f(x) = 0$ for all $x \in X$. This leads to the primal problem

$$(P^{A_g}) \quad \inf_{x \in X} (g \circ A)(x)$$

and to the perturbation function $\Phi^{A_g} : X \times Y \to \overline{\mathbb{R}}$, $\Phi^{A_g}(x, y) = g(Ax + y)$. The conjugate of Φ^{A_g}, $(\Phi^{A_g})^* : X^* \times Y^* \to \overline{\mathbb{R}}$, is, for all $(x^*, y^*) \in X^* \times Y^*$, given by

$$(\Phi^{A_g})^*(x^*, y^*) = f^*(x^* - A^* y^*) + g^*(y^*) = \begin{cases} g^*(y^*), & \text{if } x^* = A^* y^*, \\ +\infty, & \text{otherwise} \end{cases} \tag{2.3}$$

and it provides the following dual problem to (P^{A_g})

$$(D^{A_g}) \quad \sup_{\substack{y^* \in Y^* \\ A^* y^* = 0}} \{-g^*(y^*)\}.$$

By Theorem 1.1 it holds $v(D^{A_g}) \leq v(P^{A_g})$.

Denoting by $R(A) := \{Ax : x \in X\}$ the *range* of the operator A, the regularity conditions stated for (P^A) give rise to the following formulations

$$(RC_1^{A_g}) \;\Big|\; \exists x' \in A^{-1}(\text{dom}\, g) \text{ such that } g \text{ is continuous at } Ax',$$

in case X and Y are Fréchet spaces

$$(RC_2^{A_g}) \;\left|\; \begin{array}{l} X \text{ and } Y \text{ are Fréchet spaces, } g \text{ is lower semicontinuous} \\ \text{and } 0 \in \text{sqri}(\text{dom}\, g - R(A)) \end{array}\right.$$

along with its stronger versions

$$(RC_{2'}^{A_g}) \;\left|\; \begin{array}{l} X \text{ and } Y \text{ are Fréchet spaces, } g \text{ is lower semicontinuous} \\ \text{and } 0 \in \text{core}(\text{dom}\, g - R(A)) \end{array}\right.$$

and

$$(RC_{2''}^{A_g}) \quad \left| \begin{array}{l} X \text{ and } Y \text{ are Fréchet spaces, } g \text{ is lower semicontinuous} \\ \text{and } 0 \in \operatorname{int}(\operatorname{dom} g - R(A)), \end{array} \right.$$

which are in fact equivalent, while in the finite dimensional setting one has

$$(RC_3^{A_g}) \;\big|\; \dim(\operatorname{lin}(\operatorname{dom} g - R(A))) < +\infty \text{ and } \operatorname{ri}(\operatorname{dom} g) \cap R(A) \neq \emptyset.$$

Theorem 2.3. *Let $g : Y \to \overline{\mathbb{R}}$ be a proper and convex function and $A : X \to Y$ be a linear continuous operator such that $R(A) \cap \operatorname{dom} g \neq \emptyset$. If one of the regularity conditions $(RC_i^{A_g})$, $i \in \{1, 2, 3\}$, is fulfilled, then $v(P^{A_g}) = v(D^{A_g})$ and the dual has an optimal solution.*

The next pair of primal–dual problems, which we consider here, will be introduced as a special instance of (P^{A_g})–(D^{A_g}). For $m \geq 2$ let $f_i : X \to \overline{\mathbb{R}}, i = 1, \ldots, m$, be proper functions such that $\bigcap_{i=1}^m \operatorname{dom} f_i \neq \emptyset$ and take $Y = X^m := X \times \ldots \times X$, $g : X^m \to \overline{\mathbb{R}}$, $g(x^1, \ldots, x^m) = \sum_{i=1}^m f_i(x^i)$ and $A : X \to X^m$, $Ax = (x, \ldots, x)$. For this choice, the primal optimization problem (P^{A_g}) looks like

$$(P^{\Sigma}) \quad \inf_{x \in X} \left\{ \sum_{i=1}^m f_i(x) \right\}.$$

Since for all $(x^{1*}, \ldots, x^{m*}) \in X^* \times \ldots \times X^* = (X^*)^m$, $g^*(x^{1*}, \ldots, x^{m*}) = \sum_{i=1}^m f_i^*(x^{i*})$ and $A^*(x^{1*}, \ldots, x^{m*}) = \sum_{i=1}^m x^{i*}$, the dual (D^{A_g}) turns out to be

$$(D^{\Sigma}) \quad \sup_{\substack{x^{i*} \in X^*, i=1,\ldots,m, \\ \sum_{i=1}^m x^{i*} = 0}} \left\{ -\sum_{i=1}^m f_i^*(x^{i*}) \right\}.$$

For the primal–dual pair (P^{Σ})–(D^{Σ}) weak duality holds automatically.

As $\operatorname{dom} g = \prod_{i=1}^m \operatorname{dom} f_i$ and $A^{-1}(\operatorname{dom} g) = \bigcap_{i=1}^m \operatorname{dom} f_i$, the regularity condition $(RC_1^{A_g})$ is nothing else than

$$(RC_1^{\Sigma}) \;\Big|\; \exists x' \in \bigcap_{i=1}^m \operatorname{dom} f_i \text{ such that } f_i \text{ is continuous at } x', i = 1, \ldots, m.$$

For this special choice of g and A, one has $R(A) = \{(x, \ldots, x) : x \in X\}$, a set which we denote by Δ_{X^m}. Thus $(RC_i^{A_g})$, $i \in \{2, 2', 2'', 3\}$, lead to the following regularity conditions (if X is a Fréchet space, then X^m is also a Fréchet space):

$$(RC_2^{\Sigma}) \quad \left| \begin{array}{l} X \text{ is a Fréchet space, } f_i \text{ is lower semicontinuous, } i = 1, \ldots, m, \\ \text{and } 0 \in \operatorname{sqri}\left(\prod_{i=1}^m \operatorname{dom} f_i - \Delta_{X^m} \right) \end{array} \right.$$

along with its stronger versions

$$(RC_{2'}^{\Sigma}) \left| \begin{array}{l} X \text{ is a Fréchet space, } f_i \text{ is lower semicontinuous, } i = 1, \ldots, m, \\ \text{and } 0 \in \operatorname{core}\left(\prod_{i=1}^{m} \operatorname{dom} f_i - \Delta_{X^m}\right) \end{array} \right.$$

and

$$(RC_{2''}^{\Sigma}) \left| \begin{array}{l} X \text{ is a Fréchet space, } f_i \text{ is lower semicontinuous, } i = 1, \ldots, m, \\ \text{and } 0 \in \operatorname{int}\left(\prod_{i=1}^{m} \operatorname{dom} f_i - \Delta_{X^m}\right), \end{array} \right.$$

which are in fact equivalent, while in the finite dimensional case we have

$$(RC_3^{A_g}) \left| \ \dim\left(\operatorname{lin}\left(\prod_{i=1}^{m} \operatorname{dom} f_i - \Delta_{X^m}\right)\right) < +\infty \text{ and } \bigcap_{i=1}^{m} \operatorname{ri}(\operatorname{dom} f_i) \neq \emptyset. \right.$$

The strong duality theorem for the (P^{Σ}) and its conjugate dual problem (D^{Σ}) follows.

Theorem 2.4. *Let $f_i : X \to \overline{\mathbb{R}}, i = 1, \ldots, m$, be proper and convex functions such that $\bigcap_{i=1}^{m} \operatorname{dom} f_i \neq \emptyset$. If one of the regularity conditions (RC_i^{Σ}), $i \in \{1, 2, 3\}$, is fulfilled, then $v(P^{\Sigma}) = v(D^{\Sigma})$ and the dual has an optimal solution.*

Remark 2.5. The primal–dual pair (P^{Σ})–(D^{Σ}) can be also seen as a particular instance of (P^A)–(D^A) by dealing with one of the functions that appear in the sum separately from the others. Working in this way, one would obtain the same dual problem to (P^{Σ}), but different regularity conditions. By doing so, one can easily observe that instead of (RC_1^{Σ}) it is enough to assume that there exists $x' \in \bigcap_{i=1}^{m} \operatorname{dom} f_i$ such that $m - 1$ functions f_i are continuous at x'. Regarding the regularity conditions (RC_i^{Σ}), $i \in \{1, 2, 2', 2'', 3\}$, it is worth mentioning that in case $m = 2$ they coincide with the regularity conditions (RC_i^{id}), $i \in \{1, 2, 2', 2'', 3\}$, respectively, stated for the primal–dual pair (P^{id})–(D^{id}).

3 The Problem with Geometric and Cone Constraints

By considering different perturbation functions, we assign in this section to an optimization problem with geometric and cone constrains three dual problems and establish the relations between their optimal objective values. We also provide generalized interior point regularity conditions which turn out to be sufficient for having strong duality in all three situations.

Consider X and Z separated locally convex spaces, where Z is partially ordered by the nonempty convex cone C, i.e. on Z there is a partial order $\leq_C$ defined by $z \leq_C y \Leftrightarrow y - z \in C$ for $y, z \in Z$. We recall that $C \subseteq Z$ is a *cone* if for all $\lambda \geq 0$ and all $c \in C$ one has $\lambda c \in C$. Throughout this book we assume, as follows by the definition, that the considered cones always contain the origin. To Z we attach a greatest element with respect to $\leq_C$, which does not belong to Z and is denoted by ∞_C. Let $Z^\bullet = Z \cup \{\infty_C\}$. Then for any $z \in Z^\bullet$ one has $z \leq_C \infty_C$ and one can consider on $Z^\bullet$ the following operations: $z + \infty_C = \infty_C + z = \infty_C, t\infty_C = \infty_C$ for all $t \geq 0$ and $\langle z^*, \infty_C \rangle = +\infty$ for all $z^* \in C^* := \{\lambda \in Z^* : \langle \lambda, c \rangle \geq 0\ \forall c \in C\}$. The set C^* is said to be the *dual cone* of C.

Some of the notions that exist for functions with extended real values may be given for functions having their ranges in infinite dimensional spaces. For a function $g : X \to Z^\bullet$ we call *domain* of g the set $\operatorname{dom} g = \{x \in X : g(x) \in Z\}$ and we say that g is *proper* when $\operatorname{dom} g \neq \emptyset$. The function g is said to be C-*convex* if for all $x, y \in X$ and all $\lambda \in [0,1]$, $g(\lambda x + (1-\lambda)y) \leq_C \lambda g(x) + (1-\lambda)g(y)$. The C-*epigraph* of g is the set defined by $\operatorname{epi}_C g = \{(x,z) \in X \times Z : z \in g(x) + C\}$. Also in this setting one has that g is C-convex if and only if $\operatorname{epi}_C g$ is a convex set. For $z^* \in C^*$ we denote by $(z^*g) : X \to \overline{\mathbb{R}}$ the function defined by $(z^*g)(x) = \langle z^*, g(x) \rangle$. Obviously, for all $z^* \in C^*$, $\operatorname{dom}(z^*g) = \operatorname{dom} g$.

The primal problem which we investigate in this section is

$$(P^C) \ \inf_{x \in \mathcal{A}} f(x),$$
$$\mathcal{A} = \{x \in S : g(x) \in -C\}$$

where $S \subseteq X$ is a given nonempty set and $f : X \to \overline{\mathbb{R}}$ and $g : X \to Z^\bullet$ are proper functions fulfilling $\operatorname{dom} f \cap S \cap g^{-1}(-C) \neq \emptyset$. Here we denote by $g^{-1}(-C)$ the set $\{x \in X : g(x) \in -C\}$.

For the beginning, we take Z as the space of the perturbation variables and define $\Phi^{C_L} : X \times Z \to \overline{\mathbb{R}}$,

$$\Phi^{C_L}(x,z) = \begin{cases} f(x), & \text{if } x \in S, g(x) \in z - C, \\ +\infty, & \text{otherwise.} \end{cases}$$

The conjugate function of Φ^{C_L} has for all $(x^*, z^*) \in X^* \times Z^*$ the following form

$$(\Phi^{C_L})^*(x^*, z^*) = \sup_{\substack{x \in S, z \in Z \\ g(x) - z \in -C}} \{\langle x^*, x \rangle + \langle z^*, z \rangle - f(x)\} = \sup_{s \in -C} \{\langle -z^*, s \rangle\}$$

$$+ \sup_{x \in S} \{\langle x^*, x \rangle + \langle z^*, g(x) \rangle - f(x)\} = \begin{cases} (f + ((-z^*)g))_S^*(x^*), & \text{if } z^* \in -C^*, \\ +\infty, & \text{otherwise.} \end{cases} \tag{3.1}$$

The dual problem of (P^C) which we obtain by means of the perturbation function Φ^{C_L} is

$$(D^{C_L}) \quad \sup_{z^* \in C^*} \left\{-(f + (z^*g))_S^*(0)\right\},$$

or, equivalently,

$$(D^{C_L}) \quad \sup_{z^* \in C^*} \inf_{x \in S} \{f(x) + (z^*g)(x)\},$$

which is nothing else than the classical *Lagrange dual* problem to (P^C). By Theorem 1.1, we automatically have that $v(D^{C_L}) \leq v(P^C)$.

The second perturbation function we consider for (P^C) is $\Phi^{C_F} : X \times X \to \overline{\mathbb{R}}$,

$$\Phi^{C_F}(x, y) = \begin{cases} f(x + y), & \text{if } x \in S, g(x) \in -C, \\ +\infty, & \text{otherwise,} \end{cases}$$

with X being the space of the perturbation variables. As $\Phi^{C_F}(x, y) = f(x + y) + \delta_{\mathcal{A}}(x)$ for all $(x, y) \in X \times X$, its conjugate follows from (2.2), i.e.

$$(\Phi^{C_F})^*(x^*, y^*) = f^*(y^*) + \sigma_{\mathcal{A}}(x^* - y^*) \ \forall (x^*, y^*) \in X^* \times X^* \tag{3.2}$$

and this leads to the following conjugate dual problem to (P^C)

$$(D^{C_F}) \quad \sup_{y^* \in X^*} \{-f^*(y^*) - \sigma_{\mathcal{A}}(-y^*)\}.$$

Since the primal problem (P^C) can be written as

$$(P^C) \quad \inf_{x \in X} \{f(x) + \delta_{\mathcal{A}}(x)\},$$

(D^{C_F}) can be seen as the *Fenchel dual* problem to (P^C). Also in this case weak duality holds, i.e. $v(D^{C_F}) \leq v(P^C)$.

For constructing the third conjugate dual problem to (P^C), we perturb both the argument of its objective function and the cone constraints. We take $X \times Z$ as the space of perturbation variables and define $\Phi^{C_{FL}} : X \times X \times Z \to \overline{\mathbb{R}}$ by

$$\Phi^{C_{FL}}(x, y, z) = \begin{cases} f(x + y), & \text{if } x \in S, g(x) \in z - C, \\ +\infty, & \text{otherwise.} \end{cases}$$

For all $(x^*, y^*, z^*) \in X^* \times X^* \times Z^*$ the conjugate function of $\Phi^{C_{FL}}$ looks like

$$(\Phi^{C_{FL}})^*(x^*, y^*, z^*) = \sup_{\substack{x \in S, y \in X, z \in Z \\ g(x) - z \in -C}} \{\langle x^*, x\rangle + \langle y^*, y\rangle + \langle z^*, z\rangle - f(x + y)\}$$

$$= \sup_{x \in S}\{\langle x^* - y^*, x\rangle + \langle z^*, g(x)\rangle\} + \sup_{r \in X}\{\langle y^*, r\rangle - f(r)\} + \sup_{s \in -C}\{\langle -z^*, s\rangle\}$$

$$= \begin{cases} f^*(y^*) + ((-z^*)g)_S^*(x^* - y^*), & \text{if } z^* \in -C^*, \\ +\infty, & \text{otherwise,} \end{cases} \tag{3.3}$$

and it furnishes the following dual problem to (P^C)

$$(D^{C_{FL}}) \quad \sup_{y^* \in X^*, z^* \in C^*} \{-f^*(y^*) - (z^* g)_S^*(-y^*)\}.$$

We call $(D^{C_{FL}})$ the *Fenchel–Lagrange dual* problem to (P^C), since it can be seen as a combination of the classical Fenchel and Lagrange duals. By the weak duality theorem we have $v(D^{C_{FL}}) \leq v(P^C)$.

The three dual problems constructed above for (P^C) have been introduced and investigated for problems in finite dimensional spaces in [124] and for problems in infinite dimensional spaces in [39].

Remark 3.1. The name Fenchel–Lagrange for the dual problem $(D^{C_{FL}})$ can be motivated by the fact that in the definition of $\Phi^{C_{FL}}$ both the cone constraints (like for (D^{C_L})) and the argument of the objective function (like for (D^{C_F})) are perturbed. Another motivation for this name is given below.

The *Lagrange dual* problem of (P^C) can be formulated as being

$$(D^{C_L}) \quad \sup_{z^* \in C^*} \inf_{x \in X} \{f(x) + ((z^* g) + \delta_S)(x)\}.$$

For every $z^* \in C^*$ the *Fenchel dual* of the infimum problem in the objective function of the problem above looks like (cf. Section 2)

$$\sup_{y^* \in X^*} \left\{-f^*(y^*) - (z^* g)_S^*(-y^*)\right\}.$$

By replacing in the objective function of (D^{C_L}) the infimum problem by its Fenchel dual, one gets exactly the *Fenchel–Lagrange dual* of (P^C). Weak duality is automatically ensured.

Consider now the *Fenchel dual* problem of (P^C) formulated as being

$$(D^{C_F}) \quad \sup_{y^* \in X^*} \{-f^*(y^*) + \inf_{x \in \mathcal{A}} \langle y^*, x \rangle\}.$$

For every $y^* \in X^*$ the *Lagrange dual* of the infimum problem which appears in the objective function of (D^{C_F}) is

$$\sup_{z^* \in C^*} \inf_{x \in S} \{\langle y^*, x \rangle + (z^* g)(x)\} = \sup_{z^* \in C^*} \left\{-(z^* g)_S^*(-y^*)\right\}.$$

By replacing in the objective function of (D^{C_F}) the infimum problem by its Lagrange dual, what we get is again the *Fenchel–Lagrange dual* of (P^C). Weak duality is also in this situation automatically ensured. Consequently, $(D^{C_{FL}})$ can be seen as a combination of the Fenchel and Lagrange dual problems of (P^C).

The following result offers an image of the relations which exist between the optimal objective values of (D^{C_L}), (D^{C_F}) and $(D^{C_{FL}})$.

Proposition 3.2. *It holds:*

(i) $v(D^{C_{FL}}) \leq v(D^{C_L})$;
(ii) $v(D^{C_{FL}}) \leq v(D^{C_F})$.

Proof. (i) Let $z^* \in C^*$ be fixed. Since

$$\sup_{y^* \in X^*} \left\{-f^*(y^*) - (z^*g)_S^*(-y^*)\right\}$$

is the Fenchel dual of the problem

$$\inf_{x \in X} \{f(x) + ((z^*g) + \delta_S)(x)\},$$

by Theorem 1.1, it follows that

$$\sup_{y^* \in X^*} \left\{-f^*(y^*) - (z^*g)_S^*(-y^*)\right\} \le \inf_{x \in X} \{f(x) + ((z^*g) + \delta_S)(x)\}.$$

Taking the supremum over $z^* \in C^*$ in both sides of the equality above, the relation $v(D^{C_{FL}}) \le v(D^{C_L})$ follows automatically.

(ii) Let $y^* \in X^*$ be fixed. Since

$$\sup_{z^* \in C^*} \left\{-(z^*g)_S^*(-y^*)\right\}$$

is the Lagrange dual problem of

$$\inf_{x \in \mathcal{A}} \langle y^*, x\rangle,$$

using again Theorem 1.1 it yields

$$-f^*(y^*) + \sup_{z^* \in C^*} \left\{-(z^*g)_S^*(-y^*)\right\} \le -f^*(y^*) + \inf_{x \in \mathcal{A}} \langle y^*, x\rangle.$$

Taking the supremum over $y^* \in X^*$ in both sides of the inequality above, we obtain $v(D^{C_{FL}}) \le v(D^{C_F})$. □

Combining Proposition 3.2 with the fact that weak duality holds for (P^C) and the three conjugate duals, one obtains the following scheme (see also [14, 39, 124])

$$v(D^{C_{FL}}) \le \begin{array}{c} v(D^{C_L}) \\ v(D^{C_F}) \end{array} \le v(P^C). \tag{3.4}$$

The reader is referred to [124] for examples which show that in general the inequalities in (3.4) can be strict and that between $v(D^{C_L})$ and $v(D^{C_F})$ no ordering relation can be established (see also [36, 37]).

Starting from the generalized interior point regularity conditions given in Section 1, we deduce, in the following, corresponding regularity conditions and formulate corresponding strong duality results for (P^C) and the three duals treated above. To this end, we additionally assume that $S \subseteq X$ is a convex set, $f : X \to \overline{\mathbb{R}}$ is a convex function and $g : X \to Z^\bullet$ is a C-convex function. Under these hypotheses, the three perturbation functions are proper and convex and 0 is an element in the projection of their domains on the space of the perturbation variables.

At the beginning we deal with the Lagrange dual problem (D^{C_L}). The regularity condition (RC_1^{Φ}) states in this particular case that there exists $x' \in \operatorname{dom} f \cap S \cap g^{-1}(-C)$ such that the function $z \mapsto f(x') + \delta_S(x') + \delta_{g(x')+C}(z)$ is continuous at 0, which is the same with saying that there exists $x' \in \operatorname{dom} f \cap S \cap g^{-1}(-C)$ such that $0 \in \operatorname{int}(g(x') + C)$ or, equivalently, with asking that

$$(RC_1^{C_L}) \mid \exists x' \in \operatorname{dom} f \cap S \text{ such that } g(x') \in -\operatorname{int}(C).$$

This is nothing else than the classical *Slater constraint qualification*.

Before coming to the other regularity conditions, we give an overview on the notions that can be found in the literature for extensions of the lower semicontinuity from real-valued to vector-valued functions. We start with the notion of C-lower semicontinuity which has been introduced by Penot and Théra in [107] and then refined in [1, 57]. The function $g : X \to Z^\bullet$ is said to be *C-lower semicontinuous at $x \in X$* if for any neighborhood V of 0 and any $b \in Y$ satisfying $b \leqq_C g(x)$, there exists a neighborhood U of x in X such that $g(U) \subseteq b + V + C \cup \{\infty_C\}$. If g is C-lower semicontinuous at all $x \in X$, then we say that g is *C-lower semicontinuous*. Very close to this notion, Combari, Laghdir and Thibault introduced in [57] the notion of a *C-sequentially lower semicontinuous* function, which coincides with the C-lower semicontinuity when X and Z are metric spaces.

We mainly deal in this work with other two extensions of the lower semicontinuity, namely the star C-lower semicontinuity and the C-epi closedness. According to [86], we say that $g : X \to Z^\bullet$ is *star C-lower semicontinuous at $x \in X$* if for all $z^* \in C^*$ (z^*g) is lower semicontinuous at x. The function g is said to be *star C-lower semicontinuous* if it is star C-lower semicontinuous at all $x \in X$. The notion of a C-epi closed function was introduced by Luc in [94]. We call g *C-epi closed* if its C-epigraph $\operatorname{epi}_C g$ is a closed set. By [94, Theorem 5.9] follows that every star C-lower semicontinuous function is C-epi closed. One can easily observe that when $Z = \mathbb{R}$ and $C = \mathbb{R}_+$ the notions C-lower semicontinuity, and C-epi closedness coincide, as they collapse in this case into the classical lower semicontinuity.

An example of a C-epi closed function which is not star C-lower semicontinuous was given in [107, Example 1.2]. Nevertheless, this function fails to be C-convex. The function in the example below is both C-convex and C-epi closed, but not star C-lower semicontinuous.

Example 3.3. Consider the function

$$g : \mathbb{R} \to (\mathbb{R}^2)^\bullet = \mathbb{R}^2 \cup \{\infty_{\mathbb{R}^2_+}\},\ g(x) = \begin{cases} (\frac{1}{x}, x), & \text{if } x > 0, \\ \infty_{\mathbb{R}^2_+}, & \text{otherwise.} \end{cases}$$

It is easy to verify that g is $\mathbb{R}^2_+$-convex and $\mathbb{R}^2_+$-epi-closed, but not star $\mathbb{R}^2_+$-lower semicontinuous. For instance, for $z^* = (0,1)^T \in \mathbb{R}^2_+$ one has

$$((0,1)^T g)(x) = \begin{cases} x, & \text{if } x > 0, \\ +\infty, & \text{otherwise,} \end{cases}$$

which fails to be lower semicontinuous.

We come now to the class of regularity conditions which assume that X and Z are Fréchet spaces. One has $\Pr_Z(\operatorname{dom}\Phi^{C_L}) = g(\operatorname{dom} f \cap S \cap \operatorname{dom} g) + C$ and in order to guarantee the lower semicontinuity for Φ^{C_L} it is enough to assume that S is closed, f is lower semicontinuous and g is C-epi closed. Under these assumptions, the epigraph of the perturbation function

$$\operatorname{epi}\Phi^{C_L} = \{(x,z,r) \in X \times Z \times \mathbb{R} : (x,r) \in \operatorname{epi} f\} \cap (S \times Z \times \mathbb{R}) \cap (\operatorname{epi}_C g \times \mathbb{R})$$

is a closed set. These lead to the following regularity condition

$$(RC_2^{C_L}) \left|\begin{array}{l} X \text{ and } Z \text{ are Fréchet spaces, } S \text{ is closed, } f \text{ is lower semicontinuous,} \\ g \text{ is } C\text{-epi closed and } 0 \in \operatorname{sqri}(g(\operatorname{dom} f \cap S \cap \operatorname{dom} g) + C) \end{array}\right.$$

along with its stronger versions

$$(RC_{2'}^{C_L}) \left|\begin{array}{l} X \text{ and } Z \text{ are Fréchet spaces, } S \text{ is closed, } f \text{ is lower semicontinuous,} \\ g \text{ is } C\text{-epi closed and } 0 \in \operatorname{core}(g(\operatorname{dom} f \cap S \cap \operatorname{dom} g) + C) \end{array}\right.$$

and

$$(RC_{2''}^{C_L}) \left|\begin{array}{l} X \text{ and } Z \text{ are Fréchet spaces, } S \text{ is closed, } f \text{ is lower semicontinuous,} \\ g \text{ is } C\text{-epi closed and } 0 \in \operatorname{int}(g(\operatorname{dom} f \cap S \cap \operatorname{dom} g) + C), \end{array}\right.$$

which are in fact equivalent.

In Banach spaces, the condition $(RC_{2'}^{C_L})$ has been considered by Rockafellar in [112], while a particular formulation of $(RC_2^{C_L})$ has been stated for linear programming problems in [87]. Mentioning that in the finite dimensional setting one can consider the following regularity condition for the pair (P^C)–(D^{C_L})

$$(RC_3^{C_L}) \left|\begin{array}{l} \dim(\operatorname{lin}(g(\operatorname{dom} f \cap S \cap \operatorname{dom} g) + C)) < +\infty \text{ and} \\ 0 \in \operatorname{ri}(g(\operatorname{dom} f \cap S \cap \operatorname{dom} g) + C), \end{array}\right.$$

we are now able to formulate the Lagrange strong duality theorem, which is a consequence of Theorem 1.7.

Theorem 3.4. *Let $S \subseteq X$ be a nonempty convex set, $f : X \to \overline{\mathbb{R}}$ a proper and convex function and $g : X \to Z^\bullet$ a proper and C-convex function such that* $\operatorname{dom} f \cap S \cap g^{-1}(-C) \neq \emptyset$. *If one of the regularity conditions $(RC_i^{C_L})$, $i \in \{1,2,3\}$, is fulfilled, then $v(P^C) = v(D^{C_L})$ and the dual has an optimal solution.*

We turn our attention now to the regularity conditions which guarantee strong duality for (P^C) and its dual (D^{C_F}). They can be derived from the ones given in Section 2 for Fenchel duality, as $\mathcal{A}$ is a nonempty convex set. Thus $(RC_i^{\text{id}}), i \in \{1, 2, 2', 2'', 3\}$, lead to the following regularity conditions

$$(RC_1^{C_F}) \Big| \; \exists x' \in \operatorname{dom} f \cap \mathcal{A} \text{ such that } f \text{ is continuous at } x',$$

in case X is a Fréchet space

$$(RC_2^{C_F}) \left| \begin{array}{l} X \text{ is a Fréchet space, } \mathcal{A} \text{ is closed, } f \text{ is lower semicontinuous} \\ \text{and } 0 \in \operatorname{sqri}(\operatorname{dom} f - \mathcal{A}) \end{array} \right.$$

along with its stronger versions

$$(RC_{2'}^{C_F}) \left| \begin{array}{l} X \text{ is a Fréchet space, } \mathcal{A} \text{ is closed, } f \text{ is lower semicontinuous} \\ \text{and } 0 \in \operatorname{core}(\operatorname{dom} f - \mathcal{A}) \end{array} \right.$$

and

$$(RC_{2''}^{C_F}) \left| \begin{array}{l} X \text{ is a Fréchet space, } \mathcal{A} \text{ is closed, } f \text{ is lower semicontinuous} \\ \text{and } 0 \in \operatorname{int}(\operatorname{dom} f - \mathcal{A}), \end{array} \right.$$

which are in fact equivalent, while in the finite dimensional case one has

$$(RC_3^{C_F}) \Big| \; \dim(\operatorname{lin}(\operatorname{dom} f - \mathcal{A})) < +\infty \text{ and } \operatorname{ri}(\operatorname{dom} f) \cap \operatorname{ri}(\mathcal{A}) \neq \emptyset.$$

Let us notice that for having $\mathcal{A}$ closed it is enough to assume that S is closed and g is C-epi closed. From Theorem 2.2 we obtain the following result.

Theorem 3.5. *Let $S \subseteq X$ be a nonempty convex set, $f : X \to \overline{\mathbb{R}}$ a proper and convex function and $g : X \to Z^\bullet$ a proper and C-convex function such that* $\operatorname{dom} f \cap S \cap g^{-1}(-C) \neq \emptyset$*. If one of the regularity conditions $(RC_i^{C_F})$, $i \in \{1, 2, 3\}$, is fulfilled, then $v(P^C) = v(D^{C_F})$ and the dual has an optimal solution.*

Next, we particularize the regularity conditions given in Section 1 by considering as perturbation function $\Phi^{C_{FL}}$. Thus, the regularity condition (RC_1^Φ) states that there exists $x' \in \operatorname{dom} f \cap S \cap g^{-1}(-C)$ such that the function $(y, z) \mapsto f(x' + y) + \delta_S(x') + \delta_{g(x')+C}(z)$ is continuous at $(0, 0)$ and can be equivalently written as

$$(RC_1^{C_{FL}}) \left| \begin{array}{l} \exists x' \in \operatorname{dom} f \cap S \text{ such that } f \text{ is continuous at } x' \\ \text{and } g(x') \in -\operatorname{int}(C). \end{array} \right.$$

For the other conditions, we need to determine the set $\operatorname{Pr}_{X \times Z}(\operatorname{dom} \Phi^{C_{FL}})$. We have

$$(y, z) \in \operatorname{Pr}_{X \times Z}(\operatorname{dom} \Phi^{C_{FL}}) \Leftrightarrow \exists x \in X \text{ such that } \Phi^{C_{FL}}(x, y, z) < +\infty$$

$$\Leftrightarrow \exists x \in S \text{ such that } x + y \in \operatorname{dom} f \text{ and } g(x) \in z - C \Leftrightarrow \exists x \in S \text{ such that}$$

$$(y, z) \in (\operatorname{dom} f - x) \times (g(x) + C) \Leftrightarrow \exists x \in S \text{ such that } (y, z) \in \operatorname{dom} f \times C$$

$$-(x, -g(x)) \Leftrightarrow (y, z) \in \text{dom}\, f \times C - \text{epi}_{(-C)}(-g) \cap (S \times Z)$$

and so

$$\text{Pr}_{X \times Z}(\text{dom}\, \Phi^{C_{FL}}) = \text{dom}\, f \times C - \text{epi}_{(-C)}(-g) \cap (S \times Z).$$

This gives rise to the following generalized interior point regularity conditions for the primal–dual pair (P^C)–$(D^{C_{FL}})$

$(RC_2^{C_{FL}})$ | X and Z are Fréchet spaces, S is closed, f is lower semicontinuous, g is C-epi closed and $0 \in \text{sqri}\left(\text{dom}\, f \times C - \text{epi}_{(-C)}(-g) \cap (S \times Z)\right)$

along with its stronger versions

$(RC_{2'}^{C_{FL}})$ | X and Z are Fréchet spaces, S is closed, f is lower semicontinuous, g is C-epi closed and $0 \in \text{core}\left(\text{dom}\, f \times C - \text{epi}_{(-C)}(-g) \cap (S \times Z)\right)$

and

$(RC_{2''}^{C_{FL}})$ | X and Z are Fréchet spaces, S is closed, f is lower semicontinuous, g is C-epi closed and $0 \in \text{int}\left(\text{dom}\, f \times C - \text{epi}_{(-C)}(-g) \cap (S \times Z)\right)$,

which are in fact equivalent, while in the finite dimensional case we have

$(RC_3^{C_{FL}})$ | $\dim\left(\text{lin}\left(\text{dom}\, f \times C - \text{epi}_{(-C)}(-g) \cap (S \times Z)\right)\right) < +\infty$ and $0 \in \text{ri}\left(\text{dom}\, f \times C - \text{epi}_{(-C)}(-g) \cap (S \times Z)\right)$.

The strong duality theorem for the pair $(P^C) - (D^{C_{FL}})$ follows from Theorem 1.7.

Theorem 3.6. *Let $S \subseteq X$ be a nonempty convex set, $f : X \to \overline{\mathbb{R}}$ a proper and convex function and $g : X \to Z^\bullet$ a proper and C-convex function such that* $\text{dom}\, f \cap S \cap g^{-1}(-C) \neq \emptyset$. *If one of the regularity conditions $(RC_i^{C_{FL}})$, $i \in \{1, 2, 3\}$, is fulfilled, then $v(P^C) = v(D^{C_{FL}})$ and the dual has an optimal solution.*

Remark 3.7. By (3.4) it follows that whenever for (P^C) and $(D^{C_{FL}})$ strong duality holds, then $v(P^C) = v(D^{C_L}) = v(D^{C_F}) = v(D^{C_{FL}})$. Moreover, if $(\bar{y}^*, \bar{z}^*) \in X^* \times C^*$ is an optimal solution of $(D^{C_{FL}})$, then $\bar{y}^* \in X^*$ is an optimal solution of (D^{C_F}) and $\bar{z}^* \in C^*$ is an optimal solution of (D^{C_L}). This means that in this situation for the pairs $(P^C) - (D^{C_L})$ and $(P^C) - (D^{C_F})$ strong duality holds, too.

We close the section by treating a particular instance of the primal problem (P^C) for which we give some weak regularity conditions guaranteeing strong duality for the primal problem and the three conjugate dual problems assigned to it. Consider $X = \mathbb{R}^n$, $Z = \mathbb{R}^m$, $C = \mathbb{R}^m_+$, $S \subseteq \mathbb{R}^n$ a nonempty convex set, $f : \mathbb{R}^n \to \overline{\mathbb{R}}$ a proper and convex function and $g : \mathbb{R}^n \to \mathbb{R}^m$, $g(x) = (g_1(x), \ldots, g_m(x))^T$, a $\mathbb{R}^m_+$-convex function, i.e. g_i is convex for $i = 1, \ldots, m$, such that $\operatorname{dom} f \cap S \cap g^{-1}(-\mathbb{R}^m_+) \neq \emptyset$. The primal problem becomes

$$(P^C_{fin}) \quad \inf_{x \in \mathcal{A}} f(x),$$
$$\mathcal{A} = \{x \in S : g(x) \leqq 0\}$$

where by "$\leqq$" we denote the partial order induced by the non-negative orthant $\mathbb{R}^m_+$ on $\mathbb{R}^m$. Further, let $L = \{i \in \{1, \ldots, m\} : g_i \text{ is affine}\}$ and $N = \{1, \ldots, m\} \setminus L$. Consider the following regularity condition (cf. [22, 34, 38, 41, 124])

$$(RC^{C_{FL}}_{fin}) \left| \begin{array}{l} \exists x' \in \operatorname{ri}(\operatorname{dom} f) \cap \operatorname{ri}(S) \text{ such that } g_i(x') \leq 0, i \in L, \\ \text{and } g_i(x') < 0, i \in N. \end{array} \right.$$

In [124] the following strong duality theorem has been proved.

Theorem 3.8. *Let $S \subseteq \mathbb{R}^n$ be a nonempty convex set, $f : \mathbb{R}^n \to \overline{\mathbb{R}}$ a proper and convex function and $g : \mathbb{R}^n \to \mathbb{R}^m$, $g(x) = (g_1(x), \ldots, g_m(x))^T$, a vector function having each component $g_i, i = 1, \ldots, m$, convex such that* $\operatorname{dom} f \cap S \cap g^{-1}(-\mathbb{R}^m_+) \neq \emptyset$. *If the regularity condition $(RC^{C_{FL}}_{fin})$ is fulfilled, then for (P^C_{fin}) and its Fenchel–Lagrange dual (denoted by $(D^{C_{FL}}_{fin})$) strong duality holds, namely $v(P^C_{fin}) = v(D^{C_{FL}}_{fin})$ and the dual has an optimal solution.*

Remark 3.9. The condition $(RC^{C_{FL}}_{fin})$ provides strong duality for the Fenchel and Lagrange dual problems of (P^C_{fin}), too. Nevertheless, for having strong duality for (P^C_{fin}) and its Fenchel dual it is enough to assume that (cf. [113, 124])

$$(RC^{C_F}_{fin}) \mid \operatorname{ri}(\operatorname{dom} f) \cap \operatorname{ri}(\mathcal{A}) \neq \emptyset$$

holds, while for having strong duality for (P^C_{fin}) and its Lagrange dual it is enough to assume that (cf. [113, 124])

$$(RC^{C_L}_{fin}) \left| \begin{array}{l} \exists x' \in \operatorname{ri}(\operatorname{dom} f \cap S) \text{ such that } g_i(x') \leq 0, i \in L, \\ \text{and } g_i(x') < 0, i \in N \end{array} \right.$$

holds.

4 The Composed Convex Optimization Problem

In this section, we employ the general approach described in the first section to the formulation of two conjugate duals to an unconstrained composed convex optimization problem. We also derive generalized interior point regularity conditions and provide strong duality theorems. As one can see in the following sections, the importance of dealing with the composed convex problem is given by the fact that several classes of optimization problems, including the ones investigated in Sections 2 and 3, can be treated as particular instances of it.

Let X and Z be separated locally convex spaces, where Z is partially ordered by the nonempty convex cone $C \subseteq Z$. Consider $f : X \to \overline{\mathbb{R}}$ a proper function, $g : Z \to \overline{\mathbb{R}}$ a proper and C-increasing function fulfilling, by convention, $g(\infty_C) = +\infty$ and $h : X \to Z^\bullet = Z \cup \{\infty_C\}$ a proper function such that $h(\operatorname{dom} f \cap \operatorname{dom} h) \cap \operatorname{dom} g \neq \emptyset$. We say that g is *C-increasing* if for $y, z \in Z$ such that $z \leq_C y$ one has $g(z) \leq g(y)$. The primal problem we deal with in this section is

$$(P^{CC}) \quad \inf_{x \in X} \{f(x) + (g \circ h)(x)\}.$$

Consider first Z as a space of perturbation variables and $\Phi^{CC_1} : X \times Z \to \overline{\mathbb{R}}$ defined by $\Phi^{CC_1}(x, z) = f(x) + g(h(x) + z)$ as perturbation function. Its conjugate $(\Phi^{CC_1})^* : X^* \times Z^* \to \overline{\mathbb{R}}$ has for all $(x^*, z^*) \in X^* \times Z^*$ the following expression

$$(\Phi^{CC_1})^*(x^*, z^*) = \sup_{x \in X, z \in Z} \{\langle x^*, x\rangle + \langle z^*, z\rangle - f(x) - g(h(x) + z)\}$$

$$= \sup_{x \in X}\{\langle x^*, x\rangle - \langle z^*, h(x)\rangle - f(x)\} + \sup_{s \in Z}\{\langle z^*, s\rangle - g(s)\}$$

$$= (f + (z^* h))^*(x^*) + g^*(z^*).$$

Since g is C-increasing, for $z^* \notin C^*$ it holds $g^*(z^*) = +\infty$. Thus for $(x^*, z^*) \in X^* \times Z^*$ we get

$$(\Phi^{CC_1})^*(x^*, z^*) = \begin{cases} g^*(z^*) + (f + (z^* h))^*(x^*), & \text{if } z^* \in C^*, \\ +\infty, & \text{otherwise.} \end{cases} \tag{4.1}$$

Therefore, the dual problem to (P^C) that we obtain by means of Φ^{CC_1} is

$$(D^{CC_1}) \quad \sup_{z^* \in C^*} \{-g^*(z^*) - (f + (z^* h))^*(0)\}.$$

By Theorem 1.1, for the pair $(P^{CC}) - (D^{CC_1})$ weak duality holds, i.e. $v(D^{CC_1}) \leq v(P^{CC})$.

For deriving a second conjugate dual to (P^{CC}), we take $X \times Z$ as space of perturbation variables and $\Phi^{CC_2} : X \times X \times Z \to \overline{\mathbb{R}}$, $\Phi^{CC_2}(x, y, z) = f(x + y) + g(h(x) + z)$ as perturbation function. Its conjugate $(\Phi^{CC_2})^* : X^* \times X^* \times Z^* \to \overline{\mathbb{R}}$ looks for all $(x^*, y^*, z^*) \in X^* \times X^* \times Z^*$ like

$$(\Phi^{CC_2})^*(x^*, y^*, z^*) = \sup_{x, y \in X, z \in Z} \{\langle x^*, x\rangle + \langle y^*, y\rangle + \langle z^*, z\rangle - f(x + y)$$

$$-g(h(x) + z)\} = \sup_{x \in X}\{\langle x^* - y^*, x\rangle - \langle z^*, h(x)\rangle\} + \sup_{s \in Z}\{\langle z^*, s\rangle - g(s)\}$$

$$+ \sup_{r \in Y}\{\langle y^*, r\rangle - f(r)\} = (z^* h)^*(x^* - y^*) + f^*(y^*) + g^*(z^*).$$

Taking again into consideration that g^* takes outside C^* the value $+\infty$ we get

$$(\Phi^{CC_2})^*(x^*, y^*, z^*) = \begin{cases} g^*(z^*) + f^*(y^*) + (z^*h)^*(x^* - y^*), & \text{if } z^* \in C^*, \\ +\infty, & \text{otherwise.} \end{cases} \tag{4.2}$$

The conjugate dual problem of (P^{CC}) obtained via the perturbation function Φ^{CC_2} is, consequently,

$$(D^{CC_2}) \quad \sup_{y^* \in X^*, z^* \in C^*} \{-g^*(z^*) - f^*(y^*) - (z^*h)^*(-y^*)\}$$

and for the pair $(P^{CC})-(D^{CC_2})$ weak duality holds, too, i.e. $v(D^{CC_2}) \leq v(P^{CC})$.

For furnishing regularity conditions for the primal–dual pairs derived above, we additionally assume that f and g are convex functions while h is a C-convex function. In these hypotheses both perturbation functions Φ^{CC_1} and Φ^{CC_2} are proper and convex and 0 is an element in the projection of their domains on the corresponding spaces of perturbation variables.

We investigate first regularity conditions for the primal–dual pair (P^{CC}) – (D^{CC_1}). The regularity condition (RC_1^{Φ}) becomes in this particular case

$$(RC_1^{CC_1}) \left| \begin{array}{l} \exists x' \in \operatorname{dom} f \cap \operatorname{dom} h \cap h^{-1}(\operatorname{dom} g) \text{ such that} \\ g \text{ is continuous at } h(x'). \end{array} \right.$$

For the regularity conditions given in case X and Z are Fréchet spaces, we have to guarantee that Φ^{CC_1} is lower semicontinuous and $0 \in \operatorname{sqri}(\operatorname{Pr}_Z(\operatorname{dom} \Phi^{CC_1}))$. One has

$$z \in \operatorname{Pr}_Z(\operatorname{dom} \Phi^{CC_1}) \Leftrightarrow \exists x \in \operatorname{dom} f \cap \operatorname{dom} h \text{ such that } z \in \operatorname{dom} g - h(x)$$

$$\Leftrightarrow z \in \operatorname{dom} g - h(\operatorname{dom} f \cap \operatorname{dom} h)$$

and, consequently, $\operatorname{Pr}_Z(\operatorname{dom} \Phi^{CC_1}) = \operatorname{dom} g - h(\operatorname{dom} f \cap \operatorname{dom} h)$.

Next, we show that whenever f and g are lower semicontinuous and h is star C-lower semicontinuous, the lower semicontinuity of Φ^{CC_1} is ensured. For all $(x, z) \in X \times Z$ it holds

$$(\Phi^{CC_1})^{**}(x, z) = \sup_{x^* \in X^*, z^* \in C^*} \{\langle x^*, x\rangle + \langle z^*, z\rangle - (f + (z^*h))^*(x^*) - g^*(z^*)\}$$

$$= \sup_{z^* \in C^*} \left\{ \langle z^*, z\rangle - g^*(z^*) + \sup_{x^* \in X^*} \{\langle x^*, x\rangle - (f + (z^*h))^*(x^*)\} \right\}$$

$$= \sup_{z^* \in C^*} \{\langle z^*, z\rangle - g^*(z^*) + (f + (z^*h))^{**}(x)\}.$$

Since for all $z^* \in C^*$ $f + (z^*h)$ is proper, convex and lower semicontinuous and g is proper, convex and lower semicontinuous, by the Fenchel–Moreau Theorem, it yields

$$(\Phi^{CC_1})^{**}(x,z) = \sup_{z^* \in C^*} \{\langle z^*, z\rangle - g^*(z^*) + f(x) + (z^*h)(x)\}$$

$$= f(x) + \sup_{z^* \in C^*} \left\{\langle z^*, h(x) + z\rangle - g^*(z^*)\right\} = f(x) + \sup_{z^* \in Z^*} \left\{\langle z^*, h(x) + z\rangle - g^*(z^*)\right\}$$

$$= f(x) + g^{**}(h(x) + z) = f(x) + g(h(x) + z) = \Phi^{CC_1}(x,z),$$

which guarantees that Φ^{CC_1} is lower semicontinuous. We can state now the following regularity condition

$$(RC_2^{CC_1}) \left| \begin{array}{l} X \text{ and } Z \text{ are Fréchet spaces, } f \text{ and } g \text{ are lower semicontinuous,} \\ h \text{ is star } C\text{-lower semicontinuous and} \\ 0 \in \operatorname{sqri}(\operatorname{dom} g - h(\operatorname{dom} f \cap \operatorname{dom} h)) \end{array} \right.$$

along with its stronger versions

$$(RC_{2'}^{CC_1}) \left| \begin{array}{l} X \text{ and } Z \text{ are Fréchet spaces, } f \text{ and } g \text{ are lower semicontinuous,} \\ h \text{ is star } C\text{-lower semicontinuous and} \\ 0 \in \operatorname{core}(\operatorname{dom} g - h(\operatorname{dom} f \cap \operatorname{dom} h)) \end{array} \right.$$

and

$$(RC_{2''}^{CC_1}) \left| \begin{array}{l} X \text{ and } Z \text{ are Fréchet spaces, } f \text{ and } g \text{ are lower semicontinuous,} \\ h \text{ is star } C\text{-lower semicontinuous and} \\ 0 \in \operatorname{int}(\operatorname{dom} g - h(\operatorname{dom} f \cap \operatorname{dom} h)), \end{array} \right.$$

which are in fact equivalent. In the finite dimensional setting one has

$$(RC_3^{CC_1}) \left| \begin{array}{l} \dim(\operatorname{lin}(\operatorname{dom} g - h(\operatorname{dom} f \cap \operatorname{dom} h))) < +\infty \text{ and} \\ \operatorname{ri}(\operatorname{dom} g) \cap \operatorname{ri}(h(\operatorname{dom} f \cap \operatorname{dom} h)) \neq \emptyset. \end{array} \right.$$

The conditions $(RC_2^{CC_1})$ and $(RC_{2'}^{CC_1})$ have been introduced in [57] but under the assumption that h is a sequentially C-lower semicontinuous function. The condition $(RC_1^{CC_1})$ is a classical one in the literature dealing with composed convex optimization problems (see, for instance, [127, Theorem 2.8.10 (i)]), while $(RC_3^{CC_1})$ has been stated for the first time in [125]. A refinement of $(RC_3^{CC_1})$ in case when X and Z are finite dimensional spaces has been given in [26] (for more details we refer also to [25, 32]). The strong duality theorem follows (cf. Theorem 1.7).

Theorem 4.1. *Let $f : X \to \overline{\mathbb{R}}$ be a proper and convex function, $g : Z \to \overline{\mathbb{R}}$ a proper convex and C-increasing function fulfilling, by convention, $g(\infty_C) = +\infty$ and $h : X \to Z^\bullet = Z \cup \{\infty_C\}$ a proper and C-convex function such that $h(\operatorname{dom} f \cap \operatorname{dom} h) \cap \operatorname{dom} g \neq \emptyset$. If one of the regularity conditions $(RC_i^{CC_1})$, $i \in \{1, 2, 3\}$, is fulfilled, then $v(P^{CC}) = v(D^{CC_1})$ and the dual has an optimal solution.*

We come now to the second primal–dual pair for which we provide regularity conditions and a strong duality result, too. The condition (RC_1^{Φ}) leads in this case to

$(RC_1^{CC_2})$ | $\exists x' \in \operatorname{dom} f \cap \operatorname{dom} h \cap h^{-1}(\operatorname{dom} g)$ such that f is continuous at x' and g is continuous at $h(x')$.

When X and Z are Fréchet spaces, as in (RC_2^{Φ}), we have to guarantee that Φ^{CC_2} is lower semicontinuous and $0 \in \operatorname{sqri}(\operatorname{Pr}_{X\times Z}(\operatorname{dom} \Phi^{CC_2}))$. Assuming that f and g are lower semicontinuous and h is star C-lower semicontinuous, then Φ^{CC_2} is lower semicontinuous, too. Further, we have

$$(y, z) \in \operatorname{Pr}_{X\times Z}(\operatorname{dom} \Phi^{CC_2})$$

$$\Leftrightarrow \exists x \in \operatorname{dom} h \text{ such that } x + y \in \operatorname{dom} f \text{ and } h(x) + z \in \operatorname{dom} g$$
$$\Leftrightarrow \exists x \in \operatorname{dom} h \text{ such that } (y, z) \in \operatorname{dom} f \times \operatorname{dom} g - (x, h(x))$$
$$\Leftrightarrow (y, z) \in \operatorname{dom} f \times \operatorname{dom} g - \operatorname{epi}_C h,$$

where in the last equivalence we used that g is C-increasing. In conclusion,

$$\operatorname{Pr}_{X\times Z}(\operatorname{dom} \Phi^{CC_2}) = \operatorname{dom} f \times \operatorname{dom} g - \operatorname{epi}_C h$$

and one obtains the following regularity condition

$(RC_2^{CC_2})$ | X and Z are Fréchet spaces, f and g are lower semicontinuous, h is star C-lower semicontinuous and $0 \in \operatorname{sqri}(\operatorname{dom} f \times \operatorname{dom} g - \operatorname{epi}_C h)$

along with its stronger versions

$(RC_{2'}^{CC_2})$ | X and Z are Fréchet spaces, f and g are lower semicontinuous, h is star C-lower semicontinuous and $0 \in \operatorname{core}(\operatorname{dom} f \times \operatorname{dom} g - \operatorname{epi}_C h)$

and

$(RC_{2''}^{CC_2})$ | X and Z are Fréchet spaces, f and g are lower semicontinuous, h is star C-lower semicontinuous and $0 \in \operatorname{int}(\operatorname{dom} f \times \operatorname{dom} g - \operatorname{epi}_C h)$,

which are in fact equivalent, while in the finite dimensional case one has

$(RC_3^{CC_2})$ | $\dim(\operatorname{lin}(\operatorname{dom} f \times \operatorname{dom} g - \operatorname{epi}_C h)) < +\infty$ and $(\operatorname{ri}(\operatorname{dom} f) \times \operatorname{ri}(\operatorname{dom} g)) \cap \operatorname{ri}(\operatorname{epi}_C h) \neq \emptyset$.

Applying again Theorem 1.7, the following strong duality result can be stated.

Theorem 4.2. *Let $f : X \to \overline{\mathbb{R}}$ be a proper and convex function, $g : Z \to \overline{\mathbb{R}}$ a proper convex and C-increasing function fulfilling, by convention, $g(\infty_C) = +\infty$ and $h : X \to Z^\bullet = Z \cup \{\infty_C\}$ a proper and C-convex function such that $h(\operatorname{dom} f \cap \operatorname{dom} h) \cap \operatorname{dom} g \neq \emptyset$. If one of the regularity conditions $(RC_i^{CC_2})$, $i \in \{1, 2, 3\}$, is fulfilled, then $v(P^{CC}) = v(D^{CC_2})$ and the dual has an optimal solution.*

Remark 4.3. Since for all $z^* \in C^*$

$$(f + (z^* h))^*(0) \leq \inf_{y^* \in X^*} \{f^*(y^*) + (z^* h)^*(-y^*)\},$$

one has in general that $v(D^{CC_2}) \leq v(D^{CC_1}) \leq v(P^{CC})$. This means that whenever for (P^{CC}) and (D^{CC_2}) strong duality holds, then we have $v(D^{CC_2}) = v(D^{CC_1}) = v(P^{CC})$. Moreover, if $(\bar{y}^*, \bar{z}^*) \in X^* \times C^*$ is an optimal solution for (D^{CC_2}), then $\bar{z}^* \in C^*$ is an optimal solution for (D^{CC_1}). Thus for (P^{CC}) and (D^{CC_1}) strong duality holds, too.

Chapter II
Moreau–Rockafellar Formulae and Closedness-Type Regularity Conditions

5 Generalized Moreau–Rockafellar Formulae

Throughout this chapter, we assume that all topological dual spaces of the separated locally convex spaces considered are endowed with the corresponding weak* topologies. In this section, we give *generalized Moreau–Rockafellar formulae* expressed via the perturbation function Φ considered in Section 1 as well as *closedness-type regularity conditions* for the general optimization problem (PG). These will be particularized in the following sections to the different classes of convex functions and corresponding convex optimization problems, respectively, introduced in the previous chapter (see also [27]).

Let X and Y be separated locally convex spaces and X^* and Y^* be their topological dual spaces, respectively. Given a function $\Phi : X \times Y \to \overline{\mathbb{R}}$, we have that Φ^* is convex and, consequently, the *infimal value* function of Φ^*, $\inf_{y^* \in Y^*} \Phi^*(\cdot, y^*) : X^* \to \overline{\mathbb{R}}$ is also convex. For a subset of X^* and a function defined on X^* the closure and the lower semicontinuous hull, respectively, in the weak* topology are denoted by $\operatorname{cl}_{\omega^*}$, while $\operatorname{cl}_{\omega^* \times \mathcal{R}}$ denotes the closure of a subset of $(X^*, \omega(X^*, X)) \times \mathbb{R}$. Here, we denote by $\mathcal{R}$ the natural topology on $\mathbb{R}$. The following theorem can be obtained from [110] and plays a determinant role in the investigations we make in this chapter.

Theorem 5.1. *Let $\Phi : X \times Y \to \overline{\mathbb{R}}$ be a proper, convex and lower semicontinuous function such that $0 \in \Pr_Y(\operatorname{dom} \Phi)$. Then for each $x^* \in X^*$ one has*

$$(\Phi(\cdot,0))^*(x^*) = \sup_{x \in X}\{\langle x^*, x\rangle - \Phi(x,0)\} = \operatorname{cl}_{\omega^*}\left(\inf_{y^* \in Y^*} \Phi^*(\cdot, y^*)\right)(x^*). \quad (5.1)$$

Proof. First, we determine the conjugate of $\inf_{y^* \in Y^*} \Phi^*(\cdot, y^*)$. For all $x \in X$ there is

$$\left(\inf_{y^* \in Y^*} \Phi^*(\cdot, y^*)\right)^*(x) = \sup_{x^* \in X^*}\left\{\langle x^*, x\rangle - \inf_{y^* \in Y^*} \Phi^*(x^*, y^*)\right\}$$

$$= \sup_{x^* \in X^*, y^* \in Y^*}\{\langle x^*, x\rangle - \Phi^*(x^*, y^*)\} = \Phi^{**}(x, 0).$$

R.I. Boţ, *Conjugate Duality in Convex Optimization*, Lecture Notes in Economics and Mathematical Systems 637, DOI 10.1007/978-3-642-04900-2_2,

Since Φ is proper, convex and lower semicontinuous, we get further that for all $x \in X$ it holds $\left(\inf_{y^* \in Y^*} \Phi^*(\cdot, y^*)\right)^*(x) = \Phi(x, 0)$. We prove that the function $\mathrm{cl}_{\omega^*}(\inf_{y^* \in Y^*} \Phi^*(\cdot, y^*))$ is proper. Assuming that it takes everywhere the value $+\infty$ it yields that its conjugate, which coincides with $\left(\inf_{y^* \in Y^*} \Phi^*(\cdot, y^*)\right)^*$, is everywhere $-\infty$. This contradicts the properness of Φ. The other possibility of $\mathrm{cl}_{\omega^*}(\inf_{y^* \in Y^*} \Phi^*(\cdot, y^*))$ to be improper is to take somewhere the value $-\infty$. By the same argument as above, one would have that for all $x \in X$ $\Phi(x, 0)$ is equal to $+\infty$. Since this contradicts the feasibility assumption $0 \in \mathrm{Pr}_Y(\mathrm{dom}(\Phi))$, $\mathrm{cl}_{\omega^*}(\inf_{y^* \in Y^*} \Phi^*(\cdot, y^*))$ is everywhere greater than $-\infty$, therefore it is proper.

The first equality in (5.1) arises from the definition of the conjugate function, while, by the Fenchel–Moreau Theorem, we obtain

$$(\Phi(\cdot, 0))^* = \left(\inf_{y^* \in Y^*} \Phi^*(\cdot, y^*)\right)^{**} = \mathrm{cl}_{\omega^*}\left(\inf_{y^* \in Y^*} \Phi^*(\cdot, y^*)\right).$$

This concludes the proof. □

A first consequence of this result follows.

Theorem 5.2. *Let $\Phi : X \times Y \to \overline{\mathbb{R}}$ be a proper, convex and lower semicontinuous function with $0 \in \mathrm{Pr}_Y(\mathrm{dom}(\Phi))$. Then*

$$\mathrm{epi}((\Phi(\cdot, 0))^*) = \mathrm{cl}_{\omega^* \times \mathcal{R}}\left(\mathrm{epi}\left(\inf_{y^* \in Y^*} \Phi^*(\cdot, y^*)\right)\right) = \mathrm{cl}_{\omega^* \times \mathcal{R}}\left(\mathrm{Pr}_{X^* \times \mathbb{R}}(\mathrm{epi}\,\Phi^*)\right). \tag{5.2}$$

Proof. Whenever $(x^*, r) \in \mathrm{Pr}_{X^* \times \mathbb{R}}(\mathrm{epi}\,\Phi^*)$ it is clear that $\inf_{y^* \in Y^*} \Phi^*(x^*, y^*) \leq r$, thus $(x^*, r) \in \mathrm{epi}\left(\inf_{y^* \in Y^*} \Phi^*(\cdot, y^*)\right)$.

If $(x^*, r) \in \mathrm{epi}\left(\inf_{y^* \in Y^*} \Phi^*(\cdot, y^*)\right)$, then for each $\varepsilon > 0$ there is an $y^* \in Y^*$ such that $\Phi^*(x^*, y^*) \leq r + \varepsilon$. Thus for all $\varepsilon > 0$ we have $(x^*, r + \varepsilon) \in \cup_{y^* \in Y^*} \mathrm{epi}(\Phi^*(\cdot, y^*)) = \mathrm{Pr}_{X^* \times \mathbb{R}}(\mathrm{epi}\,\Phi^*)$, and so $(x^*, r) \in \mathrm{cl}_{\omega^* \times \mathcal{R}}(\mathrm{Pr}_{X^* \times \mathbb{R}}(\mathrm{epi}\,\Phi^*))$. Then we get

$$\mathrm{Pr}_{X^* \times \mathbb{R}}(\mathrm{epi}\,\Phi^*) \subseteq \mathrm{epi}\left(\inf_{y^* \in Y^*} \Phi^*(\cdot, y^*)\right) \subseteq \mathrm{cl}_{\omega^* \times \mathcal{R}}\left(\mathrm{Pr}_{X^* \times \mathbb{R}}(\mathrm{epi}\,\Phi^*)\right),$$

which implies the second equality in (5.2). Since the previous theorem yields $\mathrm{epi}((\Phi(\cdot, 0))^*) = \mathrm{cl}_{\omega^* \times \mathcal{R}}\left(\mathrm{epi}\left(\inf_{y^* \in Y^*} \Phi^*(\cdot, y^*)\right)\right)$, we get the desired conclusion. □

The theorems proved above lead to the following statement, given also in [49] (see also [19]).

Corollary 5.3. *Let $\Phi : X \times Y \to \overline{\mathbb{R}}$ be a proper, convex and lower semicontinuous function such that $0 \in \mathrm{Pr}_Y(\mathrm{dom}(\Phi))$. Then $\mathrm{Pr}_{X^* \times \mathbb{R}}(\mathrm{epi}\,\Phi^*)$ is closed in $(X^*, \omega(X^*, X)) \times \mathbb{R}$ if and only if*

$$\sup_{x \in X}\{\langle x^*, x\rangle - \Phi(x,0)\} = \min_{y^* \in Y^*} \Phi^*(x^*, y^*) \ \forall x^* \in X^*. \tag{5.3}$$

As usual, we write instead of inf (sup) for an attained infimum (supremum) min (max).

In the following, we introduce the concept of *stable strong duality*. For every $x^* \in X^*$ we consider the *extension of the primal optimization problem* (PG)

$$(PG^{x^*}) \quad \inf_{x \in X}\{\Phi(x,0) - \langle x^*, x\rangle\}.$$

The function $\Phi^{x^*} : X \times Y \to \overline{\mathbb{R}}$, $\Phi^{x^*}(x,y) = \Phi(x,y) - \langle x^*, x\rangle$ is a perturbation function for (PG^{x^*}) and since

$$(\Phi^{x^*})^*(u^*, y^*) = \Phi^*(u^* + x^*, y^*) \ \forall (u^*, y^*) \in X^* \times Y^*,$$

it introduces the following conjugate dual

$$(DG^{x^*}) \quad \sup_{y^* \in Y^*}\{-(\Phi^{x^*})^*(0, y^*)\}$$

or, equivalently,

$$(DG^{x^*}) \quad \sup_{y^* \in Y^*}\{-\Phi^*(x^*, y^*)\}.$$

Definition 5.4. We say that for the optimization problems (PG) and (DG) *stable strong duality* holds, if for all $x^* \in X^*$ for (PG^{x^*}) and (DG^{x^*}) strong duality holds, this means that $v(PG^{x^*}) = v(DG^{x^*})$ and the dual (DG^{x^*}) has an optimal solution.

By Corollary 5.3 if follows that, if Φ is a proper and convex function with $0 \in \Pr_Y(\operatorname{dom}\Phi)$, asking that

$$(RC_4^{\Phi}) \left| \begin{array}{l} \Phi \text{ is lower semicontinuous and } \Pr_{X^* \times \mathbb{R}}(\operatorname{epi}\Phi^*) \text{ is closed in} \\ (X^*, \omega(X^*, X)) \times \mathbb{R} \end{array} \right.$$

is a sufficient condition for stable strong duality. Consequently, as stable strong duality implies strong duality (take in (5.3) $x^* = 0$), (RC_4^{Φ}) is a sufficient condition for strong duality, too. Because of the way in which it was formulated, we call it *closedness-type regularity condition*.

Under the assumption that Φ is proper and convex with $0 \in \Pr_Y(\operatorname{dom}\Phi)$ for all $x^* \in X^*$, we have $\operatorname{dom}\Phi^{x^*} = \operatorname{dom}\Phi$ and therefore the generalized interior point regularity conditions (RC_i^{Φ}), $i \in \{1,2,3\}$, are also sufficient for having strong duality for (PG^{x^*}) and (DG^{x^*}). Thus one can state the following result (see also Corollary 2.7.3 in [127]).

Theorem 5.5. *Let $\Phi : X \times Y \to \overline{\mathbb{R}}$ be a proper and convex function such that $0 \in \Pr_Y(\operatorname{dom}\Phi)$. If one of the regularity conditions (RC_i^{Φ}), $i \in \{1,2,3\}$, is fulfilled, then for (PG) and (DG) stable strong duality holds, which is nothing else than*

$$\sup_{x \in X}\{\langle x^*, x\rangle - \Phi(x,0)\} = \min_{y^* \in Y^*} \Phi^*(x^*, y^*) \ \forall x^* \in X^*.$$

Next, we derive by means of the considerations made above closedness-type regularity conditions and corresponding stable strong duality results for the primal–dual pairs treated in Sections 2–4. We provide also examples where the closedness-type conditions are satisfied, but the generalized interior point ones fail.

6 Stable Strong Duality for the Composed Convex Optimization Problem

In this section, we work in the same setting as in the Section 4 and consider X and Z separated locally convex spaces, where Z is partially ordered by the nonempty convex cone $C \subseteq Z$, $f : X \to \overline{\mathbb{R}}$ a proper and convex function, $g : Z \to \overline{\mathbb{R}}$ a proper, convex and C-increasing function fulfilling, by convention, $g(\infty_C) = +\infty$ and $h : X \to Z^\bullet = Z \cup \{\infty_C\}$ a proper and C-convex function such that $h(\operatorname{dom} f \cap \operatorname{dom} h) \cap \operatorname{dom} g \neq \emptyset$. We give two generalized Moreau–Rockafellar formulae for $f + g \circ h$ by using the perturbation functions Φ^{CC_1} and Φ^{CC_2} introduced in Section 4. To this end, we additionally assume that f and g are lower semicontinuous, while h is *star C-lower semicontinuous*. As we have seen, the perturbation function $\Phi^{CC_1} : X \times Z \to \overline{\mathbb{R}}$, $\Phi^{CC_1}(x,z) = f(x) + g(h(x)+z)$, which we treat first, is in these hypotheses proper, convex and lower semicontinuous. From (4.1) we get via Theorems 5.1 and 5.2 the following results, respectively.

Theorem 6.1. *One has*

$$(f + g \circ h)^* = \operatorname{cl}_{\omega^*}\left(\inf_{z^* \in C^*}\{g^*(z^*) + (f + (z^*h))^*(\cdot)\}\right). \tag{6.1}$$

Theorem 6.2. *It holds*

$$\begin{aligned} \operatorname{epi}((f + g \circ h)^*) &= \operatorname{cl}_{\omega^* \times \mathcal{R}}\left(\operatorname{epi}\left(\inf_{z^* \in C^*}\{g^*(z^*) + (f + (z^*h))^*(\cdot)\}\right)\right) \\ &= \operatorname{cl}_{\omega^* \times \mathcal{R}}\left(\bigcup_{z^* \in \operatorname{dom} g^*}\left((0, g^*(z^*)) + \operatorname{epi}(f + (z^*h))^*\right)\right). \end{aligned} \tag{6.2}$$

Proof. To get the result, we must only establish the set $\operatorname{Pr}_{X^* \times \mathbb{R}}(\operatorname{epi}(\Phi^{CC_1})^*)$. One has $(x^*, r) \in \operatorname{Pr}_{X^* \times \mathbb{R}}(\operatorname{epi}(\Phi^{CC_1})^*)$ if and only if there is some $z^* \in C^*$ such that $g^*(z^*) + (f + (z^*h))^*(x^*) \leq r$. Using that $\operatorname{dom} g^* \subseteq C^*$ this turns out to be equivalent to the existence of a $z^* \in \operatorname{dom} g^*$ fulfilling $(x^*, r) \in (0, g^*(z^*)) + \operatorname{epi}(f + (z^*h))^*$ and further to

$$(x^*, r) \in \bigcup_{z^* \in \operatorname{dom} g^*}\left((0, g^*(z^*)) + \operatorname{epi}(f + (z^*h))^*\right).$$

Consequently, (5.2) is nothing else than (6.2). □

Corollary 5.3 leads to the following stable strong duality result.

Theorem 6.3. *We have*

$$(f + g \circ h)^*(x^*) = \min_{z^* \in C^*} \{g^*(z^*) + (f + (z^*h))^*(x^*)\} \ \forall x^* \in X^* \tag{6.3}$$

if and only if

$$\bigcup_{z^* \in \operatorname{dom} g^*} ((0, g^*(z^*)) + \operatorname{epi}(f + (z^*h))^*) \text{ is closed in } (X^*, \omega(X^*, X)) \times \mathbb{R}. \tag{6.4}$$

Remark 6.4. Taking into consideration Theorem 6.3 one can state for the primal problem

$$(P^{CC}) \quad \inf_{x \in X} \{f(x) + (g \circ h)(x)\}$$

and its conjugate dual problem

$$(D^{CC_1}) \quad \sup_{z^* \in C^*} \{-g^*(z^*) - (f + (z^*h))^*(0)\}$$

the following closedness-type regularity condition

$(RC_4^{CC_1})$ | f and g are lower semicontinuous, h is star C-lower semicontinuous and $\bigcup_{z^* \in \operatorname{dom} g^*} ((0, g^*(z^*)) + \operatorname{epi}(f + (z^*h))^*)$ is closed in $(X^*, \omega(X^*, X)) \times \mathbb{R}$.

Example 6.5. By Theorem 5.5 and Theorem 6.3 one has $(RC_{2''}^{CC_1}) \Leftrightarrow (RC_{2'}^{CC_1}) \Rightarrow (RC_2^{CC_1}) \Rightarrow (RC_4^{CC_1})$. In the following we present a situation where $(RC_4^{CC_1})$ is fulfilled while $(RC_2^{CC_1})$ fails. Moreover, neither $(RC_1^{CC_1})$ nor $(RC_3^{CC_1})$ are satisfied.

Let $X = Z = \mathbb{R}$, $C = \{0\}$, $f = \delta_{\mathbb{R}_-}$, $g(x) = (1/2)x^2 + \delta_{\mathbb{R}_+}(x)$, $x \in \mathbb{R}$, and $h = \mathrm{id}_{\mathbb{R}}$. Here, we denote $\mathbb{R}_+ = [0, +\infty)$ and $\mathbb{R}_- = (-\infty, 0]$. Obviously, $\operatorname{dom} f \cap \operatorname{dom} h \cap h^{-1}(\operatorname{dom} g) = \{0\}$ and so $(RC_1^{CC_1})$ is not verified. Further, it holds $\operatorname{dom} g - h(\operatorname{dom} f \cap \operatorname{dom} h) = \mathbb{R}_+$, thus $\operatorname{cone}(\operatorname{dom} g - h(\operatorname{dom} f \cap \operatorname{dom} h)) = \mathbb{R}_+$ and, since this set is not a linear subspace, $(RC_2^{CC_1})$ (and also $(RC_{2'}^{CC_1})$ and $(RC_{2''}^{CC_1})$) are not fulfilled. As $\operatorname{ri}(\operatorname{dom} g) \cap \operatorname{ri}(h(\operatorname{dom} f \cap \operatorname{dom} h)) = \emptyset$, $(RC_3^{CC_1})$ fails, too.

On the other hand, for all $x^*, z^* \in \mathbb{R}$ it holds

$$g^*(z^*) = \begin{cases} 0, & \text{if } z^* < 0, \\ (1/2)(z^*)^2, & \text{if } z^* \geq 0, \end{cases} \text{ and } (f + (z^*h))(x^*) = \begin{cases} 0, & \text{if } x^* \leq z^*, \\ +\infty, & \text{otherwise.} \end{cases}$$

Thus for all $z^* < 0$ we have $(0, g^*(z^*)) + \operatorname{epi}(f + (z^*h))^* = [z^*, +\infty) \times \mathbb{R}_+$, while for all $z^* \geq 0$, $(0, g^*(z^*)) + \operatorname{epi}(f + (z^*h))^* = [z^*, +\infty) \times [(1/2)(z^*)^2, +\infty)$. So

$$\bigcup_{z^*\in \operatorname{dom} g^*} \left((0, g^*(z^*)) + \operatorname{epi}(f + (z^*h))^*\right) = \mathbb{R} \times \mathbb{R}_+,$$

which is a closed set and this proves that $(RC_4^{CC_1})$ holds.

The following characterization of (6.3) is valid without any convexity or topological assumption for the functions involved, in other words we only assume that $f : X \to \overline{\mathbb{R}}$ is a proper function, $g : Z \to \overline{\mathbb{R}}$ is a proper and C-increasing function fulfilling, by convention, $g(\infty_C) = +\infty$ and $h : X \to Z^\bullet = Z \cup \{\infty_C\}$ is a proper function such that $h(\operatorname{dom} f \cap \operatorname{dom} h) \cap \operatorname{dom} g \neq \emptyset$.

Theorem 6.6. *The relation in* (6.3) *is fulfilled if and only if for all $x \in X$ and $\varepsilon \geq 0$ one has*

$$\partial_\varepsilon(f + g \circ h)(x) = \bigcup_{\substack{\varepsilon_1,\varepsilon_2\geq 0,\varepsilon_1+\varepsilon_2=\varepsilon \\ z^*\in C^*\cap \partial_{\varepsilon_2} g(h(x))}} \partial_{\varepsilon_1}(f + (z^*h))(x).$$

Proof. "$\Rightarrow$" Let be $\varepsilon \geq 0$. If $x \notin \operatorname{dom} f \cap \operatorname{dom} h \cap h^{-1}(\operatorname{dom} g)$, there is nothing to be proved, as both sets are empty. In case $x \in \operatorname{dom} f \cap \operatorname{dom} h \cap h^{-1}(\operatorname{dom} g)$ the ε-subdifferential formula follows by [127, Theorem 2.8.10].

"$\Leftarrow$" Let be $x^* \in X^*$ fixed. By the properties of the conjugate function, we always have that

$$-\infty < (f + g \circ h)^*(x^*) \leq \inf_{z^*\in C^*} \{g^*(z^*) + (f + (z^*h))^*(x^*)\}.$$

In case $(f + g \circ h)^*(x^*) = +\infty$ the conclusion follows automatically. Assume that $r := (f + g \circ h)^*(x^*) \in \mathbb{R}$. Take an arbitrary $x \in \operatorname{dom} f \cap \operatorname{dom} h \cap h^{-1}(\operatorname{dom} g)$. Thus $(f + g \circ h)^*(x^*) + (f + g \circ h)(x) = \langle x^*, x\rangle + [r + (f + g \circ h)(x) - \langle x^*, x\rangle]$. Denote by $\varepsilon := r + (f + g \circ h)(x) - \langle x^*, x\rangle$. By the Young–Fenchel inequality one has $\varepsilon \geq 0$. On the other hand, by (1.3), $x^* \in \partial_\varepsilon(f + g \circ h)(x)$ and, according to the hypothesis there exist $\varepsilon_1, \varepsilon_2 \geq 0$, $\varepsilon_1 + \varepsilon_2 = \varepsilon$ and $z^* \in C^* \cap \partial_{\varepsilon_2} g(h(x))$ such that $x^* \in \partial_{\varepsilon_1}(f + (z^*h))(x)$. It yields $(f + (z^*h))^*(x^*) + (f + (z^*h))(x) \leq \langle x^*, x\rangle + \varepsilon_1$ and $g^*(z^*) + g(h(x)) \leq \langle z^*, h(x)\rangle + \varepsilon_2$ and from here

$$(f + (z^*h))^*(x^*) + g^*(z^*) + (f + g \circ h)(x) \leq \langle x^*, x\rangle + \varepsilon$$

or, equivalently,

$$(f + (z^*h))^*(x^*) + g^*(z^*) \leq r = (f + g \circ h)^*(x^*).$$

This concludes the proof. □

From Theorem 6.6 one can easily deduce that under the convexity and topological assumptions considered at the beginning of this section, the condition in (6.4) is necessary and sufficient for the existence of a ε-subdifferential sum formula for $f + g \circ h$. Moreover, any sufficient condition for (6.3) guarantees the

ε-subdifferential sum formula given in the previous theorem. This also applies for the regularity conditions $(RC_i^{CC_1})$, $i \in \{1,2,3\}$. For similar results in case $\varepsilon = 0$ we refer to the article of Combari, Laghdir and Thibault [57].

Coming to the perturbation function $\Phi^{CC_2} : X \times X \times Z \to \overline{\mathbb{R}}$, $\Phi^{CC_2}(x,y,z) = f(x+y) + g(h(x)+z)$, we recall first that in the hypotheses considered at the beginning of the section this function is proper, convex and lower semicontinuous. By means of it, we give further Moreau–Rockafellar formulae for $f + g \circ h$, provable by using Theorem 5.1 as well as other characterizations of the epigraph of the conjugate of this function which follow from Theorem 5.2. To this end, the expression of $(\Phi^{CC_2})^*$ given in (4.2) is used.

We recall first that for $k_i : X \to \overline{\mathbb{R}}, i = 1,..,m$, proper functions we denote by $k_1 \square \ldots \square k_m : X \to \overline{\mathbb{R}}$ their *infimal convolution*, which is defined by $k_1 \square \ldots \square k_m(x) = \inf\{\sum_{i=1}^m k_i(x^i) : \sum_{i=1}^m x^i = x\}$. If for $x \in X$ this infimum is attained, we say that $k_1 \square \ldots \square k_m$ is *exact at* x. If $k_1 \square \ldots \square k_m$ is exact at all $x \in X$, we say that $k_1 \square \ldots \square k_m$ is *exact*.

Theorem 6.7. *One has*

$$
\begin{aligned}
(f + g \circ h)^* &= \mathrm{cl}_{\omega^*} \left(\inf_{\substack{y^* \in X^* \\ z^* \in C^*}} \{g^*(z^*) + f^*(y^*) + (z^*h)^*(\cdot - y^*)\} \right) \\
&= \mathrm{cl}_{\omega^*} \left(\inf_{z^* \in C^*} \{g^*(z^*) + f^* \square (z^*h)^*(\cdot)\} \right).
\end{aligned} \tag{6.5}
$$

Theorem 6.8. *It holds*

$$
\begin{aligned}
\mathrm{epi}(f + g \circ h)^* &= \mathrm{cl}_{\omega^* \times \mathcal{R}} \left(\mathrm{epi} \left(\inf_{\substack{y^* \in X^* \\ z^* \in C^*}} \{g^*(z^*) + f^*(y^*) + (z^*h)^*(\cdot - y^*)\} \right) \right) \\
&= \mathrm{cl}_{\omega^* \times \mathcal{R}} \left(\mathrm{epi}\, f^* + \bigcup_{z^* \in \mathrm{dom}\, g^*} \left((0, g^*(z^*)) + \mathrm{epi}(z^*h)^* \right) \right).
\end{aligned} \tag{6.6}
$$

Proof. The first equality corresponds to the first one given in relation (5.2). For the second one we have to determine the set $\mathrm{Pr}_{X^* \times \mathbb{R}}(\mathrm{epi}(\Phi^{CC_2})^*)$. One has $(x^*, r) \in \mathrm{Pr}_{X^* \times \mathbb{R}}(\mathrm{epi}(\Phi^{CC_2})^*)$ if and only if there are some $y^* \in X^*$ and $z^* \in C^*$ such that $g^*(z^*) + f^*(y^*) + (z^*h)^*(x^* - y^*) \leq r$ or, equivalently, there exist $y^* \in X^*$ and $z^* \in C^*$ such that $(x^*, r) \in (y^*, f^*(y^*)) + (0, g^*(z^*)) + \mathrm{epi}(z^*h)^*$. Using that $\mathrm{dom}\, g^* \subseteq C^*$, this turns out to be equivalent to the existence of a $z^* \in \mathrm{dom}\, g^*$ fulfilling $(x^*, r) \in \mathrm{epi}\, f^* + (0, g^*(z^*)) + \mathrm{epi}(z^*h)^*$ and further to

$$
(x^*, r) \in \mathrm{epi}\, f^* + \bigcup_{z^* \in \mathrm{dom}\, g^*} \left((0, g^*(z^*)) + \mathrm{epi}(z^*h)^* \right).
$$

This leads to the desired result. $\square$

By Corollary 5.3 we obtain the following characterization for the conjugate of $f + g \circ h$.

Theorem 6.9. *We have*

$$(f+g\circ h)^*(x^*) = \min_{y^*\in X^*, z^*\in C^*} \{g^*(z^*) + f^*(y^*) + (z^*h)^*(x^* - y^*)\} \ \forall x^* \in X^* \tag{6.7}$$

if and only if

$$\operatorname{epi} f^* + \bigcup_{z^*\in \operatorname{dom} g^*} ((0, g^*(z^*)) + \operatorname{epi}(z^*h)^*) \text{ is closed in } (X^*, \omega(X^*, X)) \times \mathbb{R}. \tag{6.8}$$

Remark 6.10. By Theorem 6.9 it yields that for the primal problem (P^{CC}) and its conjugate dual

$$(D^{CC_2}) \qquad \sup_{y^*\in X^*, z^*\in C^*} \{-g^*(z^*) - f^*(y^*) - (z^*h)^*(-y^*)\}$$

one can formulate the following closedness-type regularity condition

$(RC_4^{CC_2})$	f and g are lower semicontinuous, h is star C-lower semicontinuous and $\operatorname{epi} f^* + \bigcup_{z^*\in \operatorname{dom} g^*} ((0, g^*(z^*)) + \operatorname{epi}(z^*h)^*)$ is closed in $(X^*, \omega(X^*, X)) \times \mathbb{R}$.

By Theorems 5.5 and 6.9 we have $(RC_{2''}^{CC_2}) \Leftrightarrow (RC_{2'}^{CC_2}) \Rightarrow (RC_2^{CC_2}) \Rightarrow (RC_4^{CC_2})$ and it is enough to consider the problem treated in Example 6.5 to see that one can have $(RC_4^{CC_2})$ fulfilled, while $(RC_2^{CC_2})$, but also $(RC_1^{CC_2})$ and $(RC_3^{CC_2})$ fail.

Indeed, as $\operatorname{dom} f \cap \operatorname{dom} h \cap h^{-1}(\operatorname{dom} g) = \{0\}$, it is obvious that $(RC_1^{CC_2})$ is not satisfied. Taking into consideration that $\operatorname{epi}_C h = \Delta_{\mathbb{R}^2} = \{(x,x) : x \in \mathbb{R}\}$ one has $\operatorname{dom} f \times \operatorname{dom} g - \operatorname{epi}_C h = \cup_{x\in\mathbb{R}}((-\infty, x] \times [x, +\infty))$ and from here one has that $0 \in \operatorname{sqri}(\operatorname{dom} f \times \operatorname{dom} g - \operatorname{epi}_C h)$ if and only if $\cup_{x\in\mathbb{R}}((-\infty, x] \times [x, +\infty)) = \{(x,y)^T \in \mathbb{R}^2 : y \geq x\}$ is a closed linear subspace. This is not the case and therefore $(RC_2^{CC_2})$ is not fulfilled. The same applies also for $(RC_{2'}^{CC_2})$ and $(RC_{2''}^{CC_2})$.

On the other hand, $\operatorname{epi} f^* + \cup_{z^*\in \operatorname{dom} g^*} ((0, g^*(z^*)) + \operatorname{epi}(z^*h)^*) = \mathbb{R}_+ \times \mathbb{R}_+ + \mathbb{R}_- \times \mathbb{R}_+ \cup \bigcup_{z^*\geq 0}\{z^*\} \times [(1/2)(z^*)^2, +\infty) = \mathbb{R} \times \mathbb{R}_+$ and this is a closed set. Consequently, $(RC_4^{CC_2})$ is valid.

The proof of the following result can be given in analogy to the one of Theorem 6.6 and this is the reason why we omit it (see [33] for more details). One should notice that like in the mentioned result we do not make any convexity or topological assumption for the functions involved.

Theorem 6.11. *The relation in* (6.7) *is fulfilled if and only if for all* $x \in X$ *and* $\varepsilon \geq 0$ *one has*

$$\partial_\varepsilon(f + g \circ h)(x) = \bigcup_{\substack{\varepsilon_1,\varepsilon_2,\varepsilon_3 \geq 0, \varepsilon_1+\varepsilon_2+\varepsilon_3=\varepsilon \\ z^* \in C^* \cap \partial_{\varepsilon_3} g(h(x))}} \left(\partial_{\varepsilon_1} f(x) + \partial_{\varepsilon_2}(z^*h))(x)\right).$$

From Theorem 6.11 follows that every sufficient condition for (6.7), like the conditions $(RC_i^{CC_2}), i \in \{1,2,3\}$, guarantees the ε-subdifferential sum formula above. For similar results in case $\varepsilon = 0$ we refer again to [57].

Remark 6.12. Since $\operatorname{epi} f^* + \cup_{z^* \in \operatorname{dom} g^*}((0, g^*(z^*)) + \operatorname{epi}(z^*h)^*)$ is contained in the set $\cup_{z^* \in \operatorname{dom} g^*}((0, g^*(z^*)) + \operatorname{epi}(f + (z^*h))^*)$ and these sets have the same closure in $(X^*, \omega(X^*, X)) \times \mathbb{R}$ (cf. Theorems 6.2 and 6.8), it follows that in case $(RC_4^{CC_2})$ is fulfilled, $(RC_4^{CC_1})$ is fulfilled, too. The following example shows that $(RC_4^{CC_1})$ is indeed weaker than $(RC_4^{CC_2})$.

Example 6.13. Let $X = \mathbb{R}^2$, $Z = \mathbb{R}$, $C = \mathbb{R}_+$ and the convex and closed sets $U = \{(x_1,x_2)^T \in \mathbb{R}^2 : x_1 \geq 0\}$ and $V = \{(x_1,x_2)^T \in \mathbb{R}^2 : 2x_1 + x_2^2 \leq 0\}$. Take $f = \delta_U$, $g = \mathrm{id}_{\mathbb{R}}$ and $h = \delta_V$. As $\operatorname{dom} g = \mathbb{R}$ it follows $\operatorname{cone}(\operatorname{dom} g - h(\operatorname{dom} f \cap \operatorname{dom} h)) = \mathbb{R}$, which is clearly a closed linear subspace. Thus $(RC_2^{CC_1})$ stands and so $(RC_4^{CC_1})$ is valid, too. Let us see whether $(RC_4^{CC_2})$ is satisfied or not in this situation. We have for $(y_1^*, y_2^*) \in \mathbb{R}^2$ and $z^* \in \mathbb{R}_+$

$$f^*(y_1^*, y_2^*) = \begin{cases} 0, & \text{if } y_1^* \leq 0, y_2^* = 0, \\ +\infty, & \text{otherwise,} \end{cases} \quad g^*(z^*) = \begin{cases} 0, & \text{if } z^* = 1, \\ +\infty, & \text{otherwise,} \end{cases}$$

and

$$h^*(y_1^*, y_2^*) = \begin{cases} \frac{(y_2^*)^2}{y_1^*}, & \text{if } y_1^* > 0, \\ 0, & \text{if } y_1^* = y_2^* = 0, \\ +\infty, & \text{otherwise.} \end{cases}$$

For $(x_1^*, x_2^*)^T \in \mathbb{R}^2$ it holds

$$(f + g \circ h)^*(x_1^*, x_2^*) = \sup_{x_1,x_2 \in \mathbb{R}} \{x_1^* x_1 + x_2^* x_2 - \delta_{U \cap V}(x_1,x_2)\} = 0,$$

while at $(x_1^*, x_2^*)^T = (1,1)^T$ we have

$$\inf_{(y_1^*,y_2^*) \in \mathbb{R}^2, z^* \in \mathbb{R}_+} \{g^*(z^*) + f^*(y_1^*, y_2^*) + (z^*h)^*(1 - y_1^*, 1 - y_2^*)\} =$$

$$\inf_{y_i^* \leq 0} h^*(1 - y_1^*, 1) = \inf_{y_1^* \leq 0} \frac{1}{1 - y_1^*} = 0,$$

but there is no $y_1^* \leq 0$ where this value is attained. This means that (6.7) is not fulfilled and so, by Theorem 6.9, $(RC_4^{CC_2})$ is not valid.

7 Stable Strong Duality for the Problem Having the Composition with a Linear Continuous Operator in the Objective Function

Consider X and Y separated locally convex spaces, the proper, convex and lower semicontinuous functions $f : X \to \overline{\mathbb{R}}$ and $g : Y \to \overline{\mathbb{R}}$ and the linear continuous operator $A : X \to Y$ fulfilling $A(\operatorname{dom} f) \cap \operatorname{dom} g \neq \emptyset$. Taking $Z = Y, C = \{0\}$ and $h : X \to Y, h(x) = Ax$, we see that we are in a special case of the general setting from Section 6. In the following, we adapt the results provided above to this particular instance.

First we notice that $C^* = Y^*$ and for any $y^* \in Y^*$ and $x^* \in X^*$ one has $(f + (y^*A))^*(x^*) = f^*(x^* - A^*y^*)$ and

$$(y^*A)^*(x^*) = \begin{cases} 0, & \text{if } A^*y^* = x^*, \\ +\infty, & \text{otherwise.} \end{cases}$$

For a proper function $k : X \to \overline{\mathbb{R}}$ and a continuous linear operator $A : X \to Y$, we define the *infimal value function of f through A* as being the function $Af : Y \to \overline{\mathbb{R}}$, $Af(y) = \inf\{f(x) : x \in X, Ax = y\}$. The formulae given in (6.1) and (6.5) collapse into the following result.

Theorem 7.1. *One has*

$$(f + g \circ A)^* = \operatorname{cl}_{\omega^*}\left(\inf_{y^* \in Y^*}\{f^*(\cdot - A^*y^*) + g^*(y^*)\}\right) = \operatorname{cl}_{\omega^*}(f^* \Box A^*g^*). \tag{7.1}$$

Proof. The first equality follows directly from Theorem 6.1 (or Theorem 6.7). The second equality is a consequence of the first equality in (6.5), which yields

$$(f + g \circ A)^* = \operatorname{cl}_{\omega^*}\left(\inf_{y^* \in X^*, z^* \in Y^*}\{g^*(z^*) + f^*(y^*) + (z^*A)^*(\cdot - y^*)\}\right).$$

For each $x^* \in X^*$ we have

$$\begin{aligned} &\inf_{\substack{y^* \in X^* \\ z^* \in Y^*}}\{g^*(z^*) + f^*(y^*) + (z^*A)^*(x^* - y^*)\} \\ &= \inf_{y^* \in X^*}\left\{f^*(y^*) + \inf_{\substack{z^* \in Y^* \\ A^*z^* = x^* - y^*}} g^*(z^*)\right\} \\ &= \inf_{y^* \in X^*}\{f^*(y^*) + A^*g^*(x^* - y^*)\} = (f^* \Box A^*g^*)(x^*), \end{aligned}$$

which leads to the desired conclusion. □

In the following statement we denote by $(A^* \times \mathrm{id}_{\mathbb{R}})(\mathrm{epi}\, g^*)$ the image of the set $\mathrm{epi}\, g^*$ through the function $(A^* \times \mathrm{id}_{\mathbb{R}}) : Y^* \times \mathbb{R} \to X^* \times \mathbb{R}$ defined by $(A^* \times \mathrm{id}_{\mathbb{R}})(y^*, r) = (A^* y^*, r)$.

Theorem 7.2. *It holds*

$$\mathrm{epi}(f + g \circ A)^* = \mathrm{cl}_{\omega^* \times \mathcal{R}} \left(\mathrm{epi} \left(\inf_{y^* \in Y^*} (g^*(y^*) + f^*(\cdot - A^* y^*)) \right) \right)$$

$$= \mathrm{cl}_{\omega^* \times \mathcal{R}} \left(\mathrm{epi}(f^* \Box A^* g^*) \right) = \mathrm{cl}_{\omega^* \times \mathcal{R}} \left(\mathrm{epi}\, f^* + (A^* \times \mathrm{id}_{\mathbb{R}})(\mathrm{epi}\, g^*) \right). \quad (7.2)$$

Proof. The first two equalities follow directly from (7.1), while the last one is a consequence of the second equality in Theorem 6.2 (or Theorem 6.8), by noticing that

$$\bigcup_{z^* \in \mathrm{dom}\, g^*} \left((0, g^*(z^*)) + \mathrm{epi}(f + (z^* A))^* \right) = \mathrm{epi}\, f^*$$

$$+ \bigcup_{z^* \in \mathrm{dom}\, g^*} \left((0, g^*(z^*)) + \mathrm{epi}(z^* A)^* \right) = \mathrm{epi}\, f^* + \bigcup_{\substack{z^* \in \mathrm{dom}\, g^* \\ r \in \mathbb{R}, r \geq g^*(z^*)}} (A^* z^*, r)$$

$$= \mathrm{epi}\, f^* + (A^* \times \mathrm{id}_{\mathbb{R}})(\mathrm{epi}\, g^*).$$

□

Theorem 6.3 (or Theorem 6.9) give rise to the following characterization of the conjugate of $f + g \circ A$ (see also [40]).

Theorem 7.3. *We have*

$$(f + g \circ A)^*(x^*) = \min_{y^* \in Y^*} \{ f^*(x^* - A^* y^*) + g^*(y^*) \} \ \forall x^* \in X^* \quad (7.3)$$

if and only if

$$\mathrm{epi}\, f^* + (A^* \times \mathrm{id}_{\mathbb{R}})(\mathrm{epi}\, g^*) \textit{ is closed in } (X^*, \omega(X^*, X)) \times \mathbb{R}. \quad (7.4)$$

Remark 7.4. For the primal–dual pair (P^A)–(D^A) (notice that in this special case both dual problems (D^{CC_1}) and (D^{CC_2}) become (D^A)), where

$$(P^A) \quad \inf_{x \in X} \{ f(x) + (g \circ A)(x) \}$$

and

$$(D^A) \quad \sup_{y^* \in Y^*} \{ -f^*(-A^* y^*) - g^*(y^*) \},$$

one can formulate the following closedness-type regularity condition (cf. [40])

$$(RC_4^A) \left| \begin{array}{l} f \text{ and } g \text{ are lower semicontinuous and } \mathrm{epi}\, f^* + (A^* \times \mathrm{id}_{\mathbb{R}})(\mathrm{epi}\, g^*) \\ \text{is closed in } (X^*, \omega(X^*, X)) \times \mathbb{R}, \end{array} \right.$$

which is weaker than (RC_i^A), $i \in \{2, 2', 2''\}$.

From Theorem 6.6 (or Theorem 6.11) we obtain the following assertion (see [33]), which has as consequence the fact that any sufficient condition for (7.3) guarantees the ε-subdifferential formula for $f + g \circ A$, fact which was also pointed out in [78, 127].

Theorem 7.5. *The relation in* (7.3) *is fulfilled if and only if for all* $x \in X$ *and* $\varepsilon \geq 0$ *one has*

$$\partial_\varepsilon (f + g \circ A)(x) = \bigcup_{\substack{\varepsilon_1, \varepsilon_2 \geq 0 \\ \varepsilon_1 + \varepsilon_2 = \varepsilon}} \left(\partial_{\varepsilon_1} f(x) + A^* \partial_{\varepsilon_2} g(Ax) \right).$$

Next, we take $X = Y$ and assume that A is the identity operator on X, while $f, g : X \to \overline{\mathbb{R}}$ are proper, convex and lower semicontinuous functions such that $\operatorname{dom} f \cap \operatorname{dom} g \neq \emptyset$. From Theorem 7.1 and Theorem 7.2 we obtain the following results (see [40, 122, 127]), the first of them being known as the classical *Moreau–Rockafellar formula*.

Theorem 7.6. *One has*

(i) $(f + g)^* = \operatorname{cl}_{\omega^*}(f^* \Box g^*)$;
(ii) $\operatorname{epi}(f + g)^* = \operatorname{cl}_{\omega^* \times \mathcal{R}}(\operatorname{epi}(f^* \Box g^*)) = \operatorname{cl}_{\omega^* \times \mathcal{R}}(\operatorname{epi} f^* + \operatorname{epi} g^*)$.

Further, from Theorem 7.3 we obtain the following statement.

Theorem 7.7. *We have*

$$(f + g)^*(x^*) = \min_{y^* \in X^*} \{f^*(x^* - y^*) + g^*(y^*)\} \ \forall x^* \in X^* \tag{7.5}$$

if and only if

$$\operatorname{epi} f^* + \operatorname{epi} g^* \textit{ is closed in } (X^*, \omega(X^*, X)) \times \mathbb{R}. \tag{7.6}$$

Without making any convexity or topological assumption for f and g, namely assuming only that they are proper with $\operatorname{dom} f \cap \operatorname{dom} g \neq \emptyset$, one can easily notice that the relation (7.5), which is rewritable as $(f + g)^* = f^* \Box g^*$ and $f^* \Box g^*$ is exact, is equivalent to $\operatorname{epi}(f + g)^* = \operatorname{epi} f^* + \operatorname{epi} g^*$.

Remark 7.8. For the primal–dual pair (P^{id})–(D^{id}), where

$$(P^{\text{id}}) \quad \inf_{x \in X} \{f(x) + g(x)\}$$

and

$$(D^{\text{id}}) \quad \sup_{y^* \in X^*} \{-f^*(-y^*) - g^*(y^*)\},$$

one can consider as closedness-type regularity condition

$$(RC_4^{\text{id}}) \left| \begin{array}{l} f \text{ and } g \text{ are lower semicontinuous and } \operatorname{epi} f^* + \operatorname{epi} g^* \\ \text{is closed in } (X^*, \omega(X^*, X)) \times \mathbb{R}. \end{array} \right.$$

The condition (RC_4^{id}) is implied by (RC_i^{id}), $i \in \{2, 2', 2''\}$ and by taking, as in Example 6.5, $X = \mathbb{R}$, $f = \delta_{\mathbb{R}_-}$ and $g(x) = (1/2)x^2 + \delta_{\mathbb{R}_+}(x)$, one gets a situation where all the generalized interior point regularity conditions formulated in Section 2 for this primal–dual pair fail, but (RC_4^{id}) is valid.

In Theorem 7.9, which we state below, we rediscover a result obtained in [47] (see also [46, 48] for partial statements) in case X is a Banach space.

Theorem 7.9. *The relation in* (7.5) *is fulfilled if and only if for all $x \in X$ and $\varepsilon \geq 0$ one has*

$$\partial_\varepsilon (f+g)(x) = \bigcup_{\substack{\varepsilon_1, \varepsilon_2 \geq 0 \\ \varepsilon_1 + \varepsilon_2 = \varepsilon}} \left(\partial_{\varepsilon_1} f(x) + \partial_{\varepsilon_2} g(x) \right).$$

We return to the initial setting, with $g : Y \to \overline{\mathbb{R}}$ a proper, convex and lower semicontinuous function and $A : X \to Y$ a linear continuous operator and assume that f is identical to 0 along with $R(A) \cap \operatorname{dom} g \neq \emptyset$. Theorems 7.1 and 7.2 lead to the following result, the first assertion of which was proved in [71].

Theorem 7.10. *One has*

(i) $(g \circ A)^* = \operatorname{cl}_{\omega^*}(A^* g^*)$;
(ii) $\operatorname{epi}(g \circ A)^* = \operatorname{cl}_{\omega^* \times \mathcal{R}}(\operatorname{epi}(A^* g^*)) = \operatorname{cl}_{\omega^* \times \mathcal{R}}((A^* \times \mathrm{id}_{\mathbb{R}})(\operatorname{epi} g^*))$.

From Theorem 7.3 one can derive the following result.

Theorem 7.11. *We have*

$$(g \circ A)^*(x^*) = \min_{y^* \in Y^*, x^* = A^* y^*} \{g^*(y^*)\} \ \forall x^* \in X^* \tag{7.7}$$

if and only if

$$(A^* \times \mathrm{id}_{\mathbb{R}})(\operatorname{epi} g^*) \text{ is closed in } (X^*, \omega(X^*, X)) \times \mathbb{R}. \tag{7.8}$$

Remark 7.12. For the primal–dual pair (P^{A_g})–(D^{A_g}), where (see Section 2)

$$(P^{A_g}) \quad \inf_{x \in X} (g \circ A)(x)$$

and

$$(D^{A_g}) \quad \sup_{\substack{y^* \in Y^* \\ A^* y^* = 0}} \{-g^*(y^*)\},$$

the following closedness-type regularity condition can be considered

$$(RC_4^{A_g}) \ \left| \ \begin{array}{l} g \text{ is lower semicontinuous and } (A^* \times \mathrm{id}_{\mathbb{R}})(\operatorname{epi} g^*) \text{ is closed} \\ \text{in } (X^*, \omega(X^*, X)) \times \mathbb{R}, \end{array} \right.$$

which is weaker than $(RC_i^{A_g})$, $i \in \{2, 2', 2''\}$.

The following characterization is a consequence of Theorem 7.5.

Theorem 7.13. *The relation in* (7.7) *is fulfilled if and only if for all* $x \in X$ *and* $\varepsilon \geq 0$ *one has*

$$\partial_\varepsilon(g \circ A)(x) = A^* \partial_\varepsilon g(Ax).$$

Further, we specialize the setting above (as in Section 2) and assume that for $m \geq 2$ the functions $f_i : X \to \overline{\mathbb{R}}, i = 1, \dots, m$, are proper, convex and lower semicontinuous such that $\cap_{i=1}^m \operatorname{dom} f_i \neq \emptyset$. Taking $Y = X^m$, $g : X^m \to \overline{\mathbb{R}}$, $g(x^1, \dots, x^m) = \sum_{i=1}^m f_i(x^i)$ and $A : X \to X^m, Ax = (x, \dots, x)$, we have that g is proper, convex and lower semicontinuous with $R(A) \cap \operatorname{dom} g \neq \emptyset$. More than that $g \circ A = \sum_{i=1}^m f_i$ and for $x^* \in X^*$ $A^*g^*(x^*) = \inf\{\sum_{i=1}^m f_i^*(x_i^*) : \sum_{i=1}^m x_i^* = x^*\} = f_1^* \square \dots \square f_m^*(x^*)$. Moreover, $(x^*, r) \in (A^* \times \operatorname{id}_{\mathbb{R}})(\operatorname{epi} g^*)$ if and only if there exists $(x^{1*}, \dots, x^{m*}) \in (X^*)^m$ such that $g^*(x^{1*}, \dots, x^{m*}) \leq r$ and $A^*(x^{1*}, \dots, x^{m*}) = \sum_{i=1}^m x^{i*} = x^*$, which is equivalent to the fact that there exists $(x^{1*}, \dots, x^{m*}) \in (X^*)^m$ such that $\sum_{i=1}^m f_i^*(x^{i*}) \leq r$ and $\sum_{i=1}^m x^{i*} = x^*$ and further to $(x^*, r) \in \sum_{i=1}^m \operatorname{epi} f_i^*$. These considerations along with Theorems 7.10 and 7.11 provide the following results, respectively.

Theorem 7.14. *One has*

(i) $(\sum_{i=1}^m f_i)^* = \operatorname{cl}_{\omega^*}(f_1^* \square \dots \square f_m^*)$;

(ii) $\operatorname{epi}(\sum_{i=1}^m f_i)^* = \operatorname{cl}_{\omega^* \times \mathcal{R}}(\operatorname{epi}(f_1^* \square \dots \square f_m^*)) = \operatorname{cl}_{\omega^* \times \mathcal{R}}(\sum_{i=1}^m \operatorname{epi} f_i^*)$.

Theorem 7.15. *We have*

$$\left(\sum_{i=1}^m f_i\right)^*(x^*) = \min\left\{\sum_{i=1}^m f_i^*(x_i^*) : \sum_{i=1}^m x_i^* = x^*\right\} \quad \forall x^* \in X^* \tag{7.9}$$

if and only if

$$\sum_{i=1}^m \operatorname{epi} f_i^* \textit{ is closed in } (X^*, \omega(X^*, X)) \times \mathbb{R}. \tag{7.10}$$

Remark 7.16. For the primal–dual pair (P^Σ)–(D^Σ), where (see Section 2)

$$(P^\Sigma) \quad \inf_{x \in X} \left\{\sum_{i=1}^m f_i(x)\right\}$$

and

$$(D^\Sigma) \quad \sup_{\substack{x^{i*} \in X^*, i=1,\dots,m, \\ \sum_{i=1}^m x^{i*} = 0}} \left\{-\sum_{i=1}^m f_i^*(x^{i*})\right\}$$

one can formulate the following closedness-type regularity condition

$$(RC_4^{\Sigma}) \left| \begin{array}{l} f_i \text{ is lower semicontinuous}, i = 1,\dots,m, \text{ and } \sum\limits_{i=1}^{m} \operatorname{epi} f_i^* \text{ is} \\ \text{closed in } (X^*, \omega(X^*,X)) \times \mathbb{R}, \end{array} \right.$$

which is weaker than (RC_i^{Σ}), $i \in \{2, 2', 2''\}$.

Theorem 7.13 leads to the following result.

Theorem 7.17. *The relation in* (7.9) *is fulfilled if and only if for all* $x \in X$ *and* $\varepsilon \geq 0$ *one has*

$$\partial_{\varepsilon}\left(\sum_{i=1}^{m} f_i\right)(x) = \bigcup_{\substack{\varepsilon_i \geq 0, i=1,\dots,m, \\ \sum_{i=1}^{m} \varepsilon_i = \varepsilon}} \left(\sum_{i=1}^{m} \partial_{\varepsilon_i} f_i(x)\right).$$

We close this section by giving an application of Theorem 7.17 in connection to the so-called strong conical hull intersection property. Having $C_1,\dots,C_m$ convex and closed subsets of X with $\cap_{i=1}^m C_i \neq \emptyset$, we say that $\{C_1,\dots,C_m\}$ has the *strong conical hull intersection property (strong CHIP)*, if for all $x \in \cap_{i=1}^m C_i$

$$N_{\bigcap\limits_{i=1}^{m} C_i}(x) = \sum_{i=1}^{m} N_{C_i}(x).$$

The notion of strong CHIP has been introduced by Deutsch, Li and Ward in [63] in Hilbert spaces and has proved to be useful when dealing with best approximation problems as well as in the conjugate duality theory (cf. [4, 61–63]). Obviously, $\{C_1,\dots,C_m\}$ has the strong CHIP if and only if

$$\partial\left(\delta_{\bigcap\limits_{i=1}^{m} C_i}\right)(x) = \sum_{i=1}^{m} \partial(\delta_{C_i})(x) \ \forall x \in \bigcap_{i=1}^{m} C_i.$$

Taking this into consideration we get from Theorem 7.17 the following result.

Corollary 7.18. *Let* $C_1,\dots,C_m$ *be convex and closed subsets of* X *such that* $\cap_{i=1}^m C_i \neq \emptyset$. *If* $\sum_{i=1}^m \operatorname{epi} \sigma_{C_i}$ *is closed in* $(X^*, \omega(X^*,X)) \times \mathbb{R}$, *then* $\{C_1,\dots,C_m\}$ *has the strong CHIP.*

For $m = 2$, a similar result has been given by Burachik and Jeyakumar in [48, Theorem 3.1] for X a Banach space. A kind of reverse result of [48, Theorem 3.1] was provided in [68] for X an Euclidean space and under different supplementary geometric properties for the two convex and closed sets. On the other hand, Ng and Song have given a sufficient condition for strong CHIP providing X is a Fréchet space and a generalized interior point regularity condition is fulfilled (cf. [102, Theorem 4.3]). Corollary 7.18 improves this result to general spaces and shows that it works under a weaker regularity condition. For relations between strong CHIP and the so-called *bounded linear regularity* property which plays an important role when providing convergence rates for algorithms solving convex optimization problems, we refer to [5].

8 Stable Strong Duality for the Problem with Geometric and Cone Constraints

Consider X and Z separated locally convex spaces, where Z is partially ordered by the nonempty convex cone $C \subseteq Z$, $S \subseteq X$ a nonempty convex and closed set, $f : X \to \overline{\mathbb{R}}$ a proper, convex and lower semicontinuous function and $g : X \to Z^\bullet$ a proper, C-convex and C-epi closed function such that $\operatorname{dom} f \cap S \cap g^{-1}(-C) \neq \emptyset$. We define $h : X \to Z^\bullet$ as being

$$h(x) = \begin{cases} g(x), & \text{if } x \in S, \\ \infty_C, & \text{otherwise.} \end{cases}$$

Thus h is a proper, C-convex and C-epi closed function and for all $x \in X$ it holds

$$(f + \delta_{-C} \circ h)(x) = \begin{cases} f(x), & \text{if } x \in S, g(x) \in -C, \\ +\infty, & \text{otherwise.} \end{cases}$$

The function δ_{-C} is proper, convex, lower semicontinuous and C-increasing and $h(\operatorname{dom} f \cap \operatorname{dom} h) \cap (-C) \neq \emptyset$. The perturbation functions considered in Section 6 become $\Phi^{CC_1} : X \times Z \to \overline{\mathbb{R}}$, $\Phi^{CC_1}(x, z) = f(x) + \delta_{\{(u,v) \in S \times Z : g(u) \in -v-C\}}(x, z)$ and $\Phi^{CC_2} : X \times X \times Z \to \overline{\mathbb{R}}$, $\Phi^{CC_2}(x, y, z) = f(x+y) + \delta_{\{(u,v) \in S \times Z : g(u) \in -v-C\}}(x, z)$, and are (except for a sign in the cone constraints, which has no influence on the construction of the corresponding dual) the perturbation functions which lead to the Lagrange and Fenchel–Lagrange dual problems, respectively, of the problem

$$(P^C) \inf_{x \in \mathcal{A}} f(x).$$
$$\mathcal{A} = \{x \in S : g(x) \in -C\}$$

Both Φ^{CC_1} and Φ^{CC_2} are proper, convex and lower semicontinuous. It is worth mentioning that for guaranteeing the lower semicontinuity of the perturbation functions in this case, it is not necessary to assume that h is star C-lower semicontinuous. Further, one can apply the results given in Section 6 to this particular setting and establish further Moreau–Rockafellar formulae as well as stable strong duality statements for (P^C). Using that $\delta^*_{-C} = \sigma_{-C} = \delta_{C^*}$, Theorems 6.1 and 6.7 furnish the following statement.

Theorem 8.1. *One has for all $x^* \in X^*$*

$$\sup_{\substack{x\in S\\ g(x)\in -C}} \{\langle x^*, x\rangle - f(x)\} = (f + \delta_{\{y\in S: g(y)\in -C\}})^*(x^*)$$
$$= \mathrm{cl}_{\omega^*}\left(\inf_{z^*\in C^*}(f + (z^*g) + \delta_S)^*\right)(x^*) = \mathrm{cl}_{\omega^*}\left(\inf_{z^*\in C^*}(f^*\Box((z^*g) + \delta_S)^*)\right)(x^*). \tag{8.1}$$

Theorems 6.2 and 6.8 provide the following characterizations by means of epigraphs.

Theorem 8.2. *It holds*

$$\mathrm{epi}\,(f + \delta_{\{y\in S: g(y)\in -C\}})^* = \mathrm{cl}_{\omega^*\times\mathcal{R}}\left(\mathrm{epi}\left(\inf_{z^*\in C^*}(f + (z^*g) + \delta_S)^*\right)\right)$$
$$= \mathrm{cl}_{\omega^*\times\mathcal{R}}\left(\bigcup_{z^*\in C^*}\mathrm{epi}(f + (z^*g) + \delta_S)^*\right) = \mathrm{cl}_{\omega^*\times\mathcal{R}}\left(\mathrm{epi}\left(\inf_{z^*\in C^*}(f^*\Box((z^*g) + \delta_S)^*\right)\right) \tag{8.2}$$
$$= \mathrm{cl}_{\omega^*\times\mathcal{R}}\left(\mathrm{epi}\,f^* + \bigcup_{z^*\in C^*}\mathrm{epi}((z^*g) + \delta_S)^*\right).$$

In this particular instance Theorem 6.3 has the following formulation (see also [24]).

Theorem 8.3. *We have*

$$\inf_{\substack{x\in S\\ g(x)\in -C}} \{f(x) - \langle x^*, x\rangle\} = \max_{z^*\in C^*}\inf_{x\in S}\{f(x) + (z^*g)(x) - \langle x^*, x\rangle\}\ \forall x^* \in X^* \tag{8.3}$$

if and only if

$$\bigcup_{z^*\in C^*}\mathrm{epi}(f + (z^*g) + \delta_S)^* \textit{ is closed in } (X^*, \omega(X^*, X)) \times \mathbb{R}. \tag{8.4}$$

Remark 8.4. Inspired by Theorem 8.3, which provides a necessary and sufficient condition for stable strong duality for (P^C) and its Lagrange dual

$$(D^{C_L})\quad \sup_{z^*\in C^*}\inf_{x\in S}\{f(x) + (z^*g)(x)\},$$

one can formulate the following closedness-type regularity condition for this primal–dual pair

$$(RC_4^{C_L}) \left| \begin{array}{l} S \text{ is closed, } f \text{ is lower semicontinuous, } g \text{ is } C\text{-epi closed and} \\ \bigcup_{z^*\in C^*}\mathrm{epi}(f + (z^*g) + \delta_S)^* \text{ is closed in } (X^*, \omega(X^*, X)) \times \mathbb{R}. \end{array}\right.$$

The fact that $(RC_4^{C_L})$ is implied by $(RC_i^{C_L}), i \in \{2, 2', 2''\}$, follows by Theorem 5.5. We show in the following example that $(RC_4^{C_L})$ is, in fact, weaker than these generalized interior point regularity conditions. Moreover, $(RC_1^{C_L})$, $(RC_3^{C_L})$ and $(RC_{fin}^{C_L})$ are also not valid in this particular instance.

Example 8.5. (cf. [84, Example 2.1]) Let $X = Z = \mathbb{R}$, $C = \mathbb{R}_+$, $S = \mathbb{R}$, $f = \mathrm{id}_{\mathbb{R}}$ and $g(x) = \max\{x, 0\}$, $x \in \mathbb{R}$. That $(RC_1^{C_L})$ fails, is obvious. We have $g(S \cap \operatorname{dom} f \cap \operatorname{dom} g) + C = \mathbb{R}_+$ and since the cone generated by this set is not a linear subspace, $(RC_2^{C_L})$ (but also $(RC_{2'}^{C_L})$ and $(RC_{2''}^{C_L})$) fail, too. It is easy to see that $(RC_3^{C_L})$ and $(RC_{fin}^{C_L})$ are far from being satisfied, while the closedness-type condition $(RC_4^{C_L})$ holds. Indeed, for all $z^* \geq 0$ and $x^* \in \mathbb{R}$ one has

$$(f + (z^*g) + \delta_S)^*(x^*) = \sup_{x \in \mathbb{R}} \{\langle x^*, x\rangle - x - \max\{z^*x, 0\}\}$$

$$= \begin{cases} 0, & \text{if } 1 \leq x^* \leq z^* + 1, \\ +\infty, & \text{otherwise.} \end{cases}$$

Therefore $\cup_{z^* \geq 0} \operatorname{epi}(f + (z^*g) + \delta_S)^* = \cup_{z^* \geq 0} [1, z^* + 1] \times \mathbb{R}_+ = [1, +\infty) \times \mathbb{R}_+$ and this is a closed set.

From Theorem 6.6 one obtains the following characterization of (8.3) via a ε-subdifferential sum formula. Also here, it is worth noticing that for the result below no convexity or topological assumptions for the sets and functions involved are needed.

Theorem 8.6. *The relation in* (8.3) *is fulfilled if and only if for all* $x \in X$ *and* $\varepsilon \geq 0$ *one has*

$$\partial_\varepsilon \left(f + \delta_{\{y \in S : g(y) \in -C\}}\right)(x) = \bigcup_{\substack{\varepsilon_1, \varepsilon_2 \geq 0, \varepsilon_1 + \varepsilon_2 = \varepsilon \\ z^* \in C^*, 0 \leq (z^*g)(x) + \varepsilon_2}} \partial_{\varepsilon_1}(f + (z^*g) + \delta_S)(x).$$

The following refinement of relation (8.3) can be obtained by using Theorem 6.9 instead of Theorem 6.3 (see also [24]).

Theorem 8.7. *We have*

$$\inf_{\substack{x \in S \\ g(x) \in -C}} \{f(x) - \langle x^*, x\rangle\} = \max_{y^* \in X^*, z^* \in C^*} \left\{-f^*(y^*) - (z^*g)_S^*(x^* - y^*)\right\} \ \forall x^* \in X^* \tag{8.5}$$

if and only if

$$\operatorname{epi} f^* + \bigcup_{z^* \in C^*} \operatorname{epi}((z^*g) + \delta_S)^* \text{ is closed in } (X^*, \omega(X^*, X)) \times \mathbb{R}. \tag{8.6}$$

Remark 8.8. The previous result allows us to state another closedness-type regularity condition for (P^C), this time for ensuring strong duality with respect to its Fenchel–Lagrange dual problem

$$(D^{C_{FL}}) \qquad \sup_{y^* \in X^*, z^* \in C^*} \left\{-f^*(y^*) - (z^*g)_S^*(-y^*)\right\},$$

and this has the following formulation

$$(RC_4^{C_{FL}}) \left| \begin{array}{l} S \text{ is closed, } f \text{ is lower semicontinuous, } g \text{ is } C\text{-epi closed and} \\ \operatorname{epi} f^* + \bigcup\limits_{z^* \in C^*} \operatorname{epi}((z^*g) + \delta_S)^* \text{ is closed in } (X^*, \omega(X^*, X)) \times \mathbb{R}. \end{array} \right.$$

As follows by Theorem 5.5, this regularity condition is weaker than $(RC_i^{C_{FL}}), i \in \{2, 2', 2''\}$. For the problem considered in Example 8.5, none of the regularity conditions given in Section 3 for the primal–dual pair (P^C)–$(D^{C_{FL}})$ is satisfied, unlike $(RC_4^{C_{FL}})$. That $(RC_1^{C_{FL}})$, $(RC_3^{C_{FL}})$ and $(RC_{fin}^{C_{FL}})$ are not fulfilled, is obvious. Further, since $\operatorname{dom} f \times C - \operatorname{epi}_{-C}(-g) \cap (S \times Z) = \mathbb{R} \times \mathbb{R}_+ + (\mathbb{R}_- \times \mathbb{R}_+) \cup \{(x, y)^T \in \mathbb{R}^2 : 0 \leq x \leq y\} = \mathbb{R} \times \mathbb{R}_+$, one has that the cone generated by this set is not a linear subspace and so $(RC_2^{C_{FL}})$ (but also $(RC_{2'}^{C_{FL}})$ and $(RC_{2''}^{C_{FL}})$) are not valid. Coming now to $(RC_4^{C_{FL}})$, we see that $f^* = \delta_{\{1\}}$ and for all $z^* \geq 0$ and $x^* \in \mathbb{R}$ it holds

$$((z^*g) + \delta_S)^*(x^*) = \sup_{x \in \mathbb{R}} \{\langle x^*, x \rangle - \max\{z^*x, 0\}\}$$
$$= \begin{cases} 0, & \text{if } 0 \leq x^* \leq z^*, \\ +\infty, & \text{otherwise.} \end{cases}$$

So $\operatorname{epi} f^* + \cup_{z^* \geq 0} \operatorname{epi}((z^*g) + \delta_S)^* = \{1\} \times \mathbb{R}_+ + \cup_{z^* \geq 0} [0, z^*] \times \mathbb{R}_+ = [1, +\infty) \times \mathbb{R}_+$ and this is a closed set. Consequently, $(RC_4^{C_{FL}})$ is verified.

By applying Theorem 6.11, we can characterize (8.5) by means of a refined ε-subdifferential sum formula, without any convexity or topological assumption.

Theorem 8.9. *The relation in* (8.5) *is fulfilled if and only if for all $x \in X$ and $\varepsilon \geq 0$ one has*

$$\partial_\varepsilon \left(f + \delta_{\{y \in S : g(y) \in -C\}}\right)(x) = \bigcup_{\substack{\varepsilon_1, \varepsilon_2, \varepsilon_3 \geq 0 \\ \varepsilon_1 + \varepsilon_2 + \varepsilon_3 = \varepsilon \\ z^* \in C^*, 0 \leq (z^*g)(x) + \varepsilon_3}} \left(\partial_{\varepsilon_1} f(x) + \partial_{\varepsilon_2}((z^*g) + \delta_S)(x)\right).$$

Remark 8.10. A closedness-type regularity condition for the primal–dual pair (P^C)–(D^{C_F}), where

$$(D^{C_F}) \qquad \sup_{y^* \in X^*} \{-f^*(y^*) - \sigma_{\mathcal{A}}(-y^*)\},$$

can be obtained directly from the considerations made in Remark 7.8. This looks like

$$(RC_4^{C_F}) \left| \begin{array}{l} f \text{ is lower semicontinuous, } \mathcal{A} \text{ is closed and } \operatorname{epi} f^* + \operatorname{epi} \sigma_{\mathcal{A}} \\ \text{is closed in } (X^*, \omega(X^*, X)) \times \mathbb{R}, \end{array} \right.$$

and it is weaker than $(RC_i^{C_F}), i \in \{2, 2', 2''\}$.

Remark 8.11. Applying Theorem 8.2 for the situation when f is identical to 0, it yields (see [39] and [85] for the case when $S = X$)

$$\operatorname{epi}\delta^*_{\{y\in S:g(y)\in -C\}} = \operatorname{epi}\sigma_{\mathcal{A}} = \operatorname{cl}_{\omega^*\times\mathcal{R}}\left(\bigcup_{z^*\in C^*}\operatorname{epi}((z^*g)+\delta_S)^*\right). \tag{8.7}$$

Thus, by Theorems 7.6 and 8.2, we obtain that

$$\begin{aligned}\operatorname{epi}\left(f+\delta_{\{y\in S:g(y)\in -C\}}\right)^* &= \operatorname{cl}_{\omega^*\times\mathcal{R}}\left(\bigcup_{z^*\in C^*}\operatorname{epi}(f+(z^*g)+\delta_S)^*\right)\\ = \operatorname{cl}_{\omega^*\times\mathcal{R}}\left(\operatorname{epi}f^*+\operatorname{epi}\sigma_{\mathcal{A}}\right) &= \operatorname{cl}_{\omega^*\times\mathcal{R}}\left(\operatorname{epi}f^*+\operatorname{cl}_{\omega^*\times\mathcal{R}}\left(\bigcup_{z^*\in C^*}\operatorname{epi}((z^*g)+\delta_S)^*\right)\right)\\ &= \operatorname{cl}_{\omega^*\times\mathcal{R}}\left(\operatorname{epi}f^*+\bigcup_{z^*\in C^*}\operatorname{epi}((z^*g)+\delta_S)^*\right).\end{aligned} \tag{8.8}$$

Since

$$\operatorname{epi}\left(f+\delta_{\{y\in S:g(y)\in -C\}}\right)^* \supseteq \operatorname{epi}f^*+\operatorname{epi}\sigma_{\mathcal{A}} \supseteq \operatorname{epi}f^*+\bigcup_{z^*\in C^*}\operatorname{epi}((z^*g)+\delta_S)^*,$$

it yields that $(RC_4^{C_{FL}}) \Rightarrow (RC_4^{C_F})$.

On the other hand, since

$$\begin{aligned}\operatorname{epi}\left(f+\delta_{\{y\in S:g(y)\in -C\}}\right)^* &\supseteq \bigcup_{z^*\in C^*}\operatorname{epi}(f+(z^*g)+\delta_S)^*\\ &\supseteq \operatorname{epi}f^*+\bigcup_{z^*\in C^*}\operatorname{epi}((z^*g)+\delta_S)^*,\end{aligned}$$

one has that $(RC_4^{C_{FL}}) \Rightarrow (RC_4^{C_L})$. In the following two examples we present two situations which illustrate that $(RC_4^{C_F})$ and, respectively, $(RC_4^{C_L})$ are not necessarily equivalent to $(RC_4^{C_{FL}})$.

Example 8.12. Let $X = Z = \mathbb{R}$, $C = \mathbb{R}_+$, $S = \mathbb{R}$, $f = \operatorname{id}_{\mathbb{R}}$ and $g(x) = x^2$, $x \in \mathbb{R}$. In this setting $\mathcal{A} = \{0\}$, $\operatorname{epi}\sigma_{\mathcal{A}} = \mathbb{R}\times\mathbb{R}_+$, $\operatorname{epi}f^* = \{1\}\times\mathbb{R}_+$ and, as $\operatorname{epi}f^*+\operatorname{epi}\sigma_{\mathcal{A}} = \mathbb{R}\times\mathbb{R}_+$ is closed, $(RC_4^{C_F})$ is satisfied. On the other hand, for $z^* = 0$ we have $\operatorname{epi}(z^*g)^* = \{0\}\times\mathbb{R}_+$, while for $z^* > 0$, $\operatorname{epi}(z^*g)^* = \cup_{x^*\in\mathbb{R}}\{x^*\}\times\left[\frac{(x^*)^2}{4z^*},+\infty\right)$. From here we get

$$\begin{aligned}\operatorname{epi}f^*+\bigcup_{z^*\in C^*}\operatorname{epi}((z^*g)+\delta_S)^* &= \{1\}\times\mathbb{R}_+ + \{0\}\times\mathbb{R}_+ \cup \bigcup_{\substack{x^*\in\mathbb{R}\\ z^*>0}}\{x^*\}\times\left[\frac{(x^*)^2}{4z^*},+\infty\right)\\ &= \{1\}\times\mathbb{R}_+ + \{0\}\times\mathbb{R}_+ \cup \mathbb{R}\times(0,+\infty) = \{1\}\times\mathbb{R}_+ \cup \mathbb{R}\times(0,+\infty).\end{aligned}$$

This is not a closed set and, consequently, $(RC_4^{C_{FL}})$ is not satisfied.

Example 8.13. Let $X = \mathbb{R}^2$, $Z = \mathbb{R}$, $C = \mathbb{R}_+$, $S = \{(x_1, x_2)^T \in \mathbb{R}^2 : 2x_1 + x_2^2 \leq 0\}$, $f = \delta_{\mathbb{R}_+ \times \mathbb{R}}$ and g identical to 0. The set which occurs in the formulation of $(RC_4^{C_L})$ is $\cup_{z^* \in C^*} \operatorname{epi}(f + (z^*g) + \delta_S)^* = \operatorname{epi}(f + \delta_S)^*$ and, being closed, it follows that this regularity condition is fulfilled. On the other hand (see Example 6.13),

$$\operatorname{epi} f^* + \bigcup_{z^* \in C^*} \operatorname{epi}((z^*g) + \delta_S)^* = \operatorname{epi} f^* + \operatorname{epi} \sigma_S = \mathbb{R}_- \times \{0\} \times \mathbb{R}_+ + \{(0,0)^T\} \times \mathbb{R}_+$$

$$\cup \bigcup_{\substack{x_1^* > 0 \\ x_2^* \in \mathbb{R}}} \left(\{(x_1^*, x_2^*)^T\} \times \left[\frac{(x_2^*)^2}{x_1^*}, +\infty \right) \right).$$

The element $(0, 1, 0)^T$, which is in the closure of this set, does not belong to the set itself, which means that $\operatorname{epi} f^* + \cup_{z^* \in C^*} \operatorname{epi}((z^*g) + \delta_S)^*$ fails to be closed. Therefore $(RC_4^{C_{FL}})$ is not fulfilled.

Remark 8.14. If f is continuous at some element belonging to $\operatorname{dom} f \cap \mathcal{A}$, then, by Theorems 5.5, 7.6 and 7.7 one has for all $z^* \in C^*$ that $\operatorname{epi}(f + (z^*g) + \delta_S)^* = \operatorname{epi} f^* + \operatorname{epi}((z^*g) + \delta_S)^*$ and, consequently, $\cup_{z^* \in C^*} \operatorname{epi}(f + (z^*g) + \delta_S)^* = \operatorname{epi} f^* + \cup_{z^* \in C^*} \operatorname{epi}((z^*g) + \delta_S)^*$. Under this assumption, (8.3) is equivalent to

$$\operatorname{epi} f^* + \bigcup_{z^* \in C^*} \operatorname{epi}((z^*g) + \delta_S)^* \text{ is closed in } (X^*, \omega(X^*, X)) \times \mathbb{R}.$$

This equivalence was proved in [86, Corollary 3.1] in case X is a Fréchet space (this assumption is actually not necessary). One should notice that, whenever f is continuous at some element from $\operatorname{dom} f \cap \mathcal{A}$, the Lagrange and Fenchel–Lagrange dual problems of (P^C) coincide.

Supposing that $g : X \to Z^\bullet$ is continuous at some element belonging to $\operatorname{dom} f \cap S$, by the same arguments, one has that $\cup_{z^* \in C^*} \operatorname{epi}((z^*g) + \delta_S)^* = \cup_{z^* \in C^*} \operatorname{epi}(z^*g)^* + \operatorname{epi} \sigma_S$ and in this case (8.7) becomes (see, for instance, [39, 83, 84, 86])

$$\operatorname{epi} \delta^*_{\{y \in S : g(y) \in -C\}} = \operatorname{epi} \sigma_{\mathcal{A}} = \operatorname{cl}_{\omega^* \times \mathcal{R}} \left(\bigcup_{z^* \in C^*} \operatorname{epi}(z^*g)^* + \operatorname{epi} \sigma_S \right).$$

We always have that

$$\operatorname{epi} \delta^*_{\{y \in S : g(y) \in -C\}} \supseteq \bigcup_{z^* \in C^*} \operatorname{epi}((z^*g) + \delta_S)^* \supseteq \bigcup_{z^* \in C^*} \operatorname{epi}(z^*g)^* + \operatorname{epi} \sigma_S \quad (8.9)$$

and this means that, asking the set $\cup_{z^* \in C^*} \operatorname{epi}(z^*g)^* + \operatorname{epi} \sigma_S$ to be closed in $(X^*, \omega(X^*, X)) \times \mathbb{R}$, the inclusion relations in (8.9) turn out to be equalities. This condition has been introduced in [84] (see also [39]) and is known in the literature under the name *closed cone constraint qualification* $(CCCQ)$.

If f is continuous at some element belonging to $\operatorname{dom} f \cap \mathcal{A}$ and $(CCCQ)$ is fulfilled, then $\operatorname{epi}(f + \delta_{\{y \in S: g(y) \in -C\}})^* = \operatorname{epi} f^* + \operatorname{epi} \sigma_{\mathcal{A}} = \operatorname{epi} f^* + \cup_{z^* \in C^*} \operatorname{epi}(z^* g)^* + \operatorname{epi} \sigma_S$, which is nothing else than

$$\operatorname{epi} f^* + \cup_{z^* \in C^*} \operatorname{epi}(z^* g)^* + \operatorname{epi} \sigma_S \text{ is closed in } (X^*, \omega(X^*, X)) \times \mathbb{R}.$$

From (8.8), Theorems 7.7, 8.3 and 8.9 one has that, under this assumption, for (P^C) and the three dual problems (D^{C_L}), (D^{C_F}) and $(D^{C_{FL}})$ strong duality holds. In fact (see, for instance, [30, 66]) we have

$$\operatorname{epi} f^* + \cup_{z^* \in C^*} \operatorname{epi}(z^* g)^* + \operatorname{epi} \sigma_S \text{ is closed in } (X^*, \omega(X^*, X)) \times \mathbb{R}.$$

if and only if for all $x^* \in X^*$

$$\inf_{\substack{x \in S \\ g(x) \in -C}} \{f(x) - \langle x^*, x \rangle\} = \max_{\substack{y^*, u^* \in X^* \\ z^* \in C^*}} \{-f^*(y^*) - (z^* g)^*(u^*) - \sigma_S(x^* - u^* - y^*)\}.$$

9 Closedness Regarding a Set

The investigations made in Section 5 make clear that when characterizing the stable strong duality via a closedness-type regularity condition, one has to assume for the perturbation function properness, convexity and lower semicontinuity. But there are a lot of situations when, for example, the lower semicontinuity is missing, like is the case when dealing with $f + g \circ h$ under the assumption that h is (only) C-epi closed. In this section we propose an approach which turns out to be useful for giving necessary and sufficient conditions for stable strong duality, even in situations when Corollary 5.3 is not applicable. We also show how this approach applies to the composed convex problem in case h fails to be star C-lower semicontinuous, being only C-epi closed.

For U, V two subsets of the separated locally convex space X we say that U *is closed regarding the set* V if $U \cap V = \operatorname{cl}(U) \cap V$. Obviously, if U is closed, then U is closed regarding any subset of X. Let Y be another separated locally convex space. The following theorem extends the assertion in Corollary 5.3.

Theorem 9.1. *Let $\Phi : X \times Y \to \overline{\mathbb{R}}$ be a proper, convex and lower semicontinuous function such that $0 \in \operatorname{Pr}_Y(\operatorname{dom} \Phi)$ and V be a nonempty subset of X^*. Then the following statements are equivalent:*

(i) $(\Phi(\cdot, 0))^*(x^*) = \sup_{x \in X}\{\langle x^*, x \rangle - \Phi(x, 0)\} = \min_{y^* \in Y^*} \Phi^*(x^*, y^*)$ $\forall x^* \in V$;

(ii) $\operatorname{Pr}_{X^* \times \mathbb{R}}(\operatorname{epi} \Phi^*)$ *is closed regarding the set* $V \times \mathbb{R}$ *in* $(X^*, \omega(X^*, X)) \times \mathbb{R}$.

Proof. (i) $\Rightarrow$ (ii) Take an arbitrary $(x^*, r) \in \operatorname{cl}_{\omega^* \times \mathcal{R}} (\operatorname{Pr}_{X^* \times \mathbb{R}}(\operatorname{epi} \Phi^*)) \cap (V \times \mathbb{R})$. By applying Theorem 5.2 we obtain that $(x^*, r) \in \operatorname{epi}(\Phi(\cdot, 0))^*$ or, equivalently,

$\sup_{x\in X}\{\langle x^*, x\rangle - \Phi(x,0)\} \leq r$. By (i), the existence of $y^* \in Y^*$ such that $\Phi^*(x^*, y^*) \leq r$ is ensured, hence $(x^*, r) \in \Pr_{X^*\times\mathbb{R}}(\operatorname{epi}\Phi^*) \cap (V \times \mathbb{R})$ and (ii) is satisfied.

(ii) $\Rightarrow$ (i) For the opposite implication we take $x^* \in V$ arbitrary. By applying the Young–Fenchel inequality it yields

$$(\Phi(\cdot,0))^*(x^*) \leq \inf_{y^*\in Y^*} \Phi^*(x^*, y^*). \tag{9.1}$$

In case $(\Phi(\cdot,0))^*(x^*) = +\infty$, there is nothing to be proven. So, we suppose that $(\Phi(\cdot,0))^*(x^*) < +\infty$. Taking into consideration that $0 \in \Pr_Y(\operatorname{dom}\Phi)$, one can easily see that $(\Phi(\cdot,0))^*(x^*) \in \mathbb{R}$. By Theorem 5.2 and (ii) follows that

$$\begin{aligned}(x^*, (\Phi(\cdot,0))^*(x^*)) \in \operatorname{epi}(\Phi(\cdot,0))^* \cap (V\times\mathbb{R}) &= \operatorname{cl}_{\omega^*\times\mathcal{R}}\left(\Pr_{X^*\times\mathbb{R}}(\operatorname{epi}\Phi^*)\right) \cap (V\times\mathbb{R})\\ &= \Pr_{X^*\times\mathbb{R}}(\operatorname{epi}\Phi^*) \cap (V\times\mathbb{R}).\end{aligned}$$

Hence there exists $\bar{y}^* \in Y^*$ such that $\Phi^*(x^*, \bar{y}^*) \leq (\Phi(\cdot,0))^*(x^*)$ and, by combining this with (9.1), we obtain $(\Phi(\cdot,0))^*(x^*) = \Phi^*(x^*, \bar{y}^*) = \min_{y^*\in Y^*}\Phi^*(x^*, y^*)$. The proof is complete.

□

Corollary 5.3 can be seen as a consequence of the previous result when taking $V := X^*$. Working with Theorem 9.1 instead of Corollary 5.3 one can generalize the statements given in Theorems 6.3, 6.9, 7.3, 7.7, 7.11, 7.15, 8.3 and 8.7. We do so only for Theorem 7.15, as we need this result in the following.

Theorem 9.2. *Let $f_i : X \to \overline{\mathbb{R}}, i = 1,\dots,m$, be proper, convex and lower semicontinuous functions such that $\cap_{i=1}^m \operatorname{dom} f_i \neq \emptyset$ and V be a nonempty subset of X^*. Then the following statements are equivalent:*

(i) $\left(\sum_{i=1}^m f_i\right)^*(x^*) = \min\left\{\sum_{i=1}^m f_i^*(x_i^*) : \sum_{i=1}^m x_i^* = x^*\right\}\ \forall x^* \in V$;

(ii) $\sum_{i=1}^m \operatorname{epi} f_i^*$ *is closed regarding the set $V\times\mathbb{R}$ in $(X^*, \omega(X^*, X))\times\mathbb{R}$.*

Consider again the setting of Section 6 with X and Z separated locally convex spaces, where Z is partially ordered by the nonempty convex cone $C \subseteq Z$, $f : X \to \overline{\mathbb{R}}$ a proper, convex and lower semicontinuous function, $g : Z \to \overline{\mathbb{R}}$ a proper, convex, C-increasing and lower semicontinuous function fulfilling, by convention, $g(\infty_C) = +\infty$ and $h : X \to Z^\bullet = Z \cup \{\infty_C\}$ a proper and C-convex functions such that $h(\operatorname{dom} f \cap \operatorname{dom} h) \cap \operatorname{dom} g \neq \emptyset$. For the latter, we suppose instead of h star C-lower semicontinuous that h is only C-epi closed. This has as consequence the fact that the perturbation functions Φ^{CC_1} and Φ^{CC_2} are not necessarily lower semicontinuous. Nevertheless, we are able to give closedness-type regularity conditions expressed via the notion of *closedness regarding the set* which are necessary and sufficient for stable strong duality for $f + g\circ h$.

To this end, we define $F_1, F_2 : X \times Y \to \overline{\mathbb{R}}$ as being $F_1(x, y) = g(y)$ and $F_2(x, y) = f(x) + \delta_{\{(u,v)\in X\times Y : h(u)-v\in -C\}}(x, y)$. For all $x^* \in X^*$ it holds

$$(f+g\circ h)^*(x^*) = \sup_{x\in X}\{\langle x^*, x\rangle - (f+g\circ h)(x)\} = \sup_{\substack{x\in X, y\in Y\\ h(x)-y\in -C}} \{\langle x^*, x\rangle - f(x) - g(y)\}$$

$$= \sup_{x\in X, y\in Y} \{\langle x^*, x\rangle - F_1(x, y) - F_2(x, y)\} = (F_1 + F_2)^*(x^*, 0). \qquad (9.2)$$

The functions F_1 and F_2 are proper, convex and lower semicontinuous (notice that for the lower semicontinuity of F_2 it is enough to assume that h is C-epi closed) fulfilling $\operatorname{dom} F_1 \cap \operatorname{dom} F_2 \neq \emptyset$. By applying Theorem 9.2 in case $m = 2$ and $V = X^* \times \{0\} \subseteq X^* \times Y^*$ it yields

$$(F_1 + F_2)^*(x^*, 0) = \min_{(u^*,v^*)\in X^*\times Y^*} \{F_1^*(u^*, v^*) + F_2^*(x^* - u^*, -v^*)\} \ \forall x^* \in X^*$$

if and only if

epi F_1^* + epi F_2^* is closed regarding the set $X^* \times \{0\} \times \mathbb{R}$ in $(X^*, \omega(X^*, X)) \times (Y^*, \omega(Y^*, Y)) \times \mathbb{R}$.

For $(u^*, v^*) \in X^* \times Y^*$ we have

$$F_1^*(u^*, v^*) = \sup_{x\in X}\langle u^*, x\rangle + \sup_{y\in Y}\{\langle v^*, y\rangle - g(y)\} = \begin{cases} g^*(v^*), & \text{if } u^* = 0,\\ +\infty, & \text{otherwise}\end{cases}$$

and

$$F_2^*(u^*, v^*) = \sup_{\substack{x\in X, y\in Y,\\ h(x)-y\in -C}} \{\langle u^*, x\rangle + \langle v^*, y\rangle - f(x)\} = \sup_{x\in X, z\in -C} \{\langle u^*, x\rangle$$

$$+\langle v^*, h(x) - z\rangle - f(x)\} = \begin{cases} (f + (-v^*h))^*(u^*), & \text{if } v^* \in -C^*,\\ +\infty, & \text{otherwise.}\end{cases}$$

Thus epi $F_1^* = \{0\}\times \operatorname{epi} g^*$ and epi $F_2^* = \cup_{z^*\in C^*}\{(x^*, -z^*, r) : (x^*, r) \in \operatorname{epi}(f + (z^*h))^*\}$ and the equivalence from above leads to the following result (see also [29]).

Theorem 9.3. *We have*

$$(f + g \circ h)^*(x^*) = \min_{z^*\in C^*} \{g^*(z^*) + (f + (z^*h))^*(x^*)\} \ \forall x^* \in X^*$$

if and only if

$\{0\} \times \operatorname{epi} g^* + \bigcup_{z^*\in C^*} \{(x^*, -z^*, r) : (x^*, r) \in \operatorname{epi}(f + (z^*h))^*\}$ *is closed regarding the set* $X^* \times \{0\} \times \mathbb{R}$ *in* $(X^*, \omega(X^*, X)) \times (Y^*, \omega(Y^*, Y)) \times \mathbb{R}$. (9.3)

Remark 9.4. For the primal–dual pair $(P^{CC}) - (D^{CC_1})$, one can formulate, by taking into consideration the statement above, a further closedness-type regularity condition

$$(\overline{RC}_4^{CC_1}) \left| \begin{array}{l} f \text{ and } g \text{ are lower semicontinuous, } h \text{ is } C\text{-epi closed and} \\ \{0\} \times \operatorname{epi} g^* + \bigcup_{z^* \in C^*} \{(x^*, -z^*, r) : (x^*, r) \in \operatorname{epi}(f + (z^*h))^*\} \text{ is} \\ \text{closed regarding the set } X^* \times \{0\} \times \mathbb{R} \text{ in } (X^*, \omega(X^*, X)) \times \\ (Y^*, \omega(Y^*, Y)) \times \mathbb{R}. \end{array} \right.$$

Noticing that for the same primal–dual pair in [58] the following generalized interior point regularity condition has been considered

$$(\overline{RC}_2^{CC_1}) \left| \begin{array}{l} X \text{ and } Y \text{ are Fréchet spaces, } f \text{ and } g \text{ are lower semicontinuous,} \\ h \text{ is } C\text{-epi closed and } 0 \in \operatorname{sqri}(\operatorname{dom} g - h(\operatorname{dom} f \cap \operatorname{dom} h)). \end{array} \right.$$

By [58, Theorem 3.4] and Theorem 9.3 it follows that $(\overline{RC}_2^{CC_1}) \Rightarrow (\overline{RC}_4^{CC_1})$. In fact, if $(\overline{RC}_2^{CC_1})$ is fulfilled, then $\{0\} \times \operatorname{epi} g^* + \cup_{z^* \in C^*} \{(x^*, -z^*, r) : (x^*, r) \in \operatorname{epi}(f + (z^*h))^*\}$ is not only closed regarding the set $X^* \times \{0\} \times \mathbb{R}$, but even closed. For the problem considered in Example 6.5, one can easily notice that $(\overline{RC}_2^{CC_1})$ is not fulfilled, while $(\overline{RC}_4^{CC_1})$ is satisfied. Indeed, the set

$$\{0\} \times \operatorname{epi} g^* + \bigcup_{z^* \in C^*} \{(x^*, -z^*, r) : (x^*, r) \in \operatorname{epi}(f + (z^*h))^*\}$$

$$= \{0\} \times \mathbb{R}_- \times \mathbb{R}_+ + \bigcup_{z^* \in \mathbb{R}} [z^*, +\infty) \times \{-z^*\} \times \mathbb{R}_+ = \mathbb{R} \times \mathbb{R} \times \mathbb{R}_+$$

is closed and, consequently, closed regarding $\mathbb{R} \times \{0\} \times \mathbb{R}$.

Let us also mention that from Theorem 9.3 it follows that the statement in (9.3) is a necessary and sufficient condition for the ε-subdifferential sum formula given in Theorem 6.6. This improves and extends the assertion in [58, Corollary 3.6].

We stay in the same setting and consider the functions $G_1, G_2 : X \times Y \to \overline{\mathbb{R}}$ defined by $G_1(x, y) = f(x)$ and $G_2(x, y) = \delta_{\{(u,v) \in X \times Y : h(u) - v \in -C\}}(x, y)$. As $F_2 = G_1 + G_2$ for all $x^* \in X^*$, it holds $(f + g \circ h)^*(x^*) = (F_1 + G_1 + G_2)^*(x^*, 0)$. The functions G_1 and G_2 are proper, convex and lower semicontinuous and fulfill $\operatorname{dom} F_1 \cap \operatorname{dom} G_1 \cap \operatorname{dom} G_2 \neq \emptyset$. By applying again Theorem 9.2 we obtain

$$(F_1 + G_1 + G_2)^*(x^*, 0) = \min_{(u_1^*, v_1^*), (u_2^*, v_2^*) \in X^* \times Y^*} \{F_1^*(u_1^*, v_1^*)$$

$$+ G_1^*(u_2^*, v_2^*) + G_2^*(x^* - u_1^* - u_2^*, -v_1^* - v_2^*)\} \ \forall x^* \in X^*$$

if and only if

$$\begin{array}{l} \operatorname{epi} F_1^* + \operatorname{epi} G_1^* + \operatorname{epi} G_2^* \text{ is closed regarding the set } X^* \times \{0\} \times \mathbb{R} \text{ in} \\ (X^*, \omega(X^*, X)) \times (Y^*, \omega(Y^*, Y)) \times \mathbb{R}. \end{array}$$

For $(u^*, v^*) \in X^* \times Y^*$ it holds

$$G_1^*(u^*, v^*) = \begin{cases} f^*(u^*), & \text{if } v^* = 0, \\ +\infty, & \text{otherwise,} \end{cases}$$

and

$$G_2^*(u^*, v^*) = \begin{cases} (-v^*h)^*(u^*), & \text{if } v^* \in -C^*, \\ +\infty, & \text{otherwise} \end{cases}$$

and so $\operatorname{epi} G_1^* = \{(x^*, 0, r) : (x^*, r) \in \operatorname{epi} f^*\}$ and $\operatorname{epi} G_2^* = \cup_{z^* \in C^*} \{(x^*, -z^*, r) : (x^*, r) \in \operatorname{epi}(z^*h)^*\}$. Now we can formulate the following result (see also [29]).

Theorem 9.5. *We have*

$$(f + g \circ h)^*(x^*) = \min_{y^* \in X^*, z^* \in C^*} \{g^*(z^*) + f^*(y^*) + (z^*h)^*(x^* - y^*)\} \ \forall x^* \in X^*$$

if and only if

$$\begin{aligned} &\{0\} \times \operatorname{epi} g^* + \{(x^*, 0, r) : (x^*, r) \in \operatorname{epi} f^*\} \\ &+ \bigcup_{z^* \in C^*} \{(x^*, -z^*, r) : (x^*, r) \in \operatorname{epi}(z^*h)^*\} \textit{ is closed regarding the set} \\ &X^* \times \{0\} \times \mathbb{R} \textit{ in } (X^*, \omega(X^*, X)) \times (Y^*, \omega(Y^*, Y)) \times \mathbb{R}. \end{aligned} \tag{9.4}$$

Remark 9.6. As in Remark 9.4, one can give also for the primal–dual pair (P^{CC}) – (D^{CC_2}) a second closedness-type regularity condition, which is applicable in case h is C-epi closed,

$(\overline{RC}_4^{CC_2})$ | f and g are lower semicontinuous, h is C-epi closed and $\{0\} \times \operatorname{epi} g^* + \{(x^*, 0, r) : (x^*, r) \in \operatorname{epi} f^*\} + \bigcup_{z^* \in C^*} \{(x^*, -z^*, r) : (x^*, r) \in \operatorname{epi}(z^*h)^*\}$ is closed regarding the set $X^* \times \{0\} \times \mathbb{R}$ in $(X^*, \omega(X^*, X)) \times (Y^*, \omega(Y^*, Y)) \times \mathbb{R}$.

Since

$$\begin{aligned} \operatorname{epi}(F_1 + F_2)^* &= \operatorname{cl}_{\omega^* \times \omega^* \times \mathcal{R}}(\operatorname{epi} F_1^* + \operatorname{epi} F_2^*) = \operatorname{cl}_{\omega^* \times \omega^* \times \mathcal{R}}(\operatorname{epi} F_1^* + \operatorname{epi} G_1^* + \operatorname{epi} G_2^*) \\ &\supseteq \operatorname{epi} F_1^* + \operatorname{epi} F_2^* \supseteq \operatorname{epi} F_1^* + \operatorname{epi} G_1^* + \operatorname{epi} G_2^*, \end{aligned}$$

it follows that if $(\overline{RC}_4^{CC_2})$ is fulfilled, then $(\overline{RC}_4^{CC_1})$ is also verified. For the problem in Example 6.13, we have that $(\overline{RC}_2^{CC_1})$ holds, which is easily verifiable, and so $(\overline{RC}_4^{CC_1})$ is satisfied. On the other hand, as proved, (6.7) is not valid and the same applies, as follows from Theorem 9.5, for (9.4). Therefore, we have a situation where, unlike $(\overline{RC}_4^{CC_1})$, $(\overline{RC}_4^{CC_2})$ is not satisfied.

By the same arguments as in Remark 9.4, one has that the statement in (9.4) is a necessary and sufficient condition for the ε-subdifferential formula provided in Theorem 6.11.

Remark 9.7. When X and Y are separated locally convex spaces, $f : X \to \overline{\mathbb{R}}$ and $g : Y \to \overline{\mathbb{R}}$ are proper, convex and lower semicontinuous functions and $A : X \to Y$ is a linear continuous operator fulfilling $A(\operatorname{dom} f) \cap \operatorname{dom} g \neq \emptyset$, then by taking $Z = Y$, $C = \{0\}$ and $h : X \to Y$, $h(x) = Ax$, the statements in (9.3) and (9.4) provide equivalent characterizations for the conjugate formula of $f + g \circ A$ and collapse to (cf. [29])

$$\operatorname{epi} f^* + (A^* \times \mathrm{id}_{\mathbb{R}})(\operatorname{epi} g^*) \text{ is closed in } (X^*, \omega(X^*, X)) \times \mathbb{R},$$

which is exactly the necessary and sufficient condition introduced in Section 7 for stable strong duality for (P^A) and (D^A).

Now consider X and Z separated locally convex spaces, where Z is partially ordered by the nonempty convex cone $C \subseteq Z$, $S \subseteq X$ a nonempty convex and closed set, $f : X \to \overline{\mathbb{R}}$ a proper, convex and lower semicontinuous function and $g : X \to Z^\bullet$ a proper, C-convex and C-epi closed function such that $\operatorname{dom} f \cap S \cap g^{-1}(-C) \neq \emptyset$. Taking $h : X \to Z^\bullet$ as

$$h(x) = \begin{cases} g(x), & \text{if } x \in S, \\ \infty_C, & \text{otherwise,} \end{cases}$$

the statements in (9.3) and (9.4) furnish equivalent characterizations for the conjugate formulae of $f + \delta_{\{y \in S : g(y) \in -C\}}$ and are nothing else than (cf. [24])

$$\bigcup_{z^* \in C^*} \operatorname{epi}(f + (z^* g) + \delta_S)^* \text{ is closed in } (X^*, \omega(X^*, X)) \times \mathbb{R}$$

and

$$\operatorname{epi} f^* + \bigcup_{z^* \in C^*} \operatorname{epi}((z^* g) + \delta_S)^* \text{ is closed in } (X^*, \omega(X^*, X)) \times \mathbb{R},$$

respectively. These are exactly the necessary and sufficient conditions introduced in Section 8 for stable strong duality for (P^C) and its Lagrange and Fenchel–Lagrange dual problem, respectively.

Remark 9.8. In this section, as well as in Section 6, we equivalently characterize the relations in (6.3) and, respectively, (6.7) by some geometric conditions involving epigraphs. The discussion will be further carried on only for the formula (6.3), as for (6.7) the things work similarly.

In Section 6 we proved, under the assumption that h is star C-lower semicontinuous, that (6.3) is valid if and only if

$$\mathcal{M} := \bigcup_{z^* \in \operatorname{dom} g^*} \big((0, g^*(z^*)) + \operatorname{epi}(f + (z^* h))^*\big) \text{ is closed in } (X^*, \omega(X^*, X)) \times \mathbb{R}.$$

In this section, by relaxing the star C-lower semicontinuity for h to C-epi closedness, we equivalently characterize (6.3) by another condition, namely,

$\mathcal{N} := \{0\} \times \operatorname{epi} g^* + \bigcup_{z^* \in C^*} \{(x^*, -z^*, r) : (x^*, r) \in \operatorname{epi}(f + (z^* h))^*\}$ is closed regarding the set $X^* \times \{0\} \times \mathbb{R}$ in $(X^*, \omega(X^*, X)) \times (Y^*, \omega(Y^*, Y)) \times \mathbb{R}$.

For all $(x^*, r) \in X^* \times \mathbb{R}$ one has that $(x^*, 0, r) \in \mathcal{N} \cap (X^* \times \{0\} \times \mathbb{R})$ if and only if $(x^*, r) \in \mathcal{M}$ and thus $\mathcal{M}$ is closed if and only if $\mathcal{N} \cap (X^* \times \{0\} \times \mathbb{R})$ is closed. On the other hand, one has

$$\mathcal{N} \cap (X^* \times \{0\} \times \mathbb{R}) \subseteq \operatorname{cl}_{\omega^* \times \omega^* \times \mathcal{R}}(\mathcal{N} \cap (X^* \times \{0\} \times \mathbb{R}))$$

$$\subseteq \operatorname{cl}_{\omega^* \times \omega^* \times \mathcal{R}}(\mathcal{N}) \cap (X^* \times \{0\} \times \mathbb{R}).$$

Therefore, whenever $\mathcal{N}$ is closed regarding the set $X^* \times \{0\} \times \mathbb{R}$ in $(X^*, \omega(X^*, X)) \times (Y^*, \omega(Y^*, Y)) \times \mathbb{R}$, then one has that $\mathcal{M}$ is closed in $(X^*, \omega(X^*, X)) \times \mathbb{R}$. This means that by strengthening the initial topological assumption on the function h, by taking h star C-lower semicontinuous, we obtain that (6.3) is equivalent to a condition that is implied by the one stated in case h is C-epi closed. Thus one *loses* something by restricting the hypotheses, but there is a *gain* in the regularity condition which equivalently characterizes (6.3). Therefore it is up to the user to decide what is more important in each specific situation: weaker hypotheses on h or weaker regularity conditions.

Noticing that a similar discussion can also be made for (6.7), we state the following open problem.

Open problem 9.9 Do there exist X and Z separated locally convex spaces, where Z is partially ordered by the nonempty convex cone $C \subseteq Z$, a proper, convex and lower semicontinuous function $f : X \to \overline{\mathbb{R}}$, a proper, convex, C-increasing and lower semicontinuous function $g : Z \to \overline{\mathbb{R}}$ fulfilling, by convention, $g(\infty_C) = +\infty$ and a proper, C-convex and C-epi closed function $h : X \to Z^\bullet = Z \cup \{\infty_C\}$ such that $h(\operatorname{dom} f \cap \operatorname{dom} h) \cap \operatorname{dom} g \neq \emptyset$, which is not star C-lower semicontinuous, such that (6.3) (or (6.7)) is fulfilled, but (9.3) (or, respectively, (9.4)) fails?

We conclude the section by making some investigations on a further situation for which the notion *closed regarding a set* turns out to be useful when characterizing stable strong duality. To this end, we consider X and Y two separated locally convex spaces, $g : Y \to \overline{\mathbb{R}}$ a proper, convex and lower semicontinuous function and $\mathcal{C} : X \rightrightarrows Y$ a convex closed process. We recall that a *process* is a multifunction $\mathcal{C} : X \rightrightarrows Y$, the graph of which, $G(\mathcal{C}) = \{(x, y) \in X \times Y : y \in \mathcal{C}(x)\}$, is a cone. When $G(\mathcal{C})$ is convex or closed we say that $\mathcal{C}$ is a *convex process* or *closed process*, respectively. The adjoint of the process $\mathcal{C}$ is the process $\mathcal{C}^* : Y^* \rightrightarrows X^*$ defined as $G(\mathcal{C}^*) = \{(y^*, x^*) \in Y^* \times X^* : (-x^*, y^*) \in (G(\mathcal{C}))^*\}$. $\mathcal{C}^*$ is a convex process which is closed in $(X^*, \omega(X^*, X)) \times (Y^*, \omega(Y^*, Y))$. We also assume that $\operatorname{dom} g \cap R(\mathcal{C}) \neq \emptyset$, where $R(\mathcal{C}) = \cup_{x \in X} \mathcal{C}(x)$ is the *range* of $\mathcal{C}$.

Consider the function $\varphi : X \to \overline{\mathbb{R}}$, $\varphi(x) = \inf\{g(y) : y \in \mathcal{C}(x)\}$, which we call *infimal value function of φ through the convex process $\mathcal{C}$*. For all $x^* \in X^*$ the conjugate of this function looks like

$$\varphi^*(x^*) = \sup_{\substack{x\in X, y\in Y\\ (x,y)\in G(\mathcal{C})}} \{\langle x^*, x\rangle - g(y)\} = \sup_{x\in X, y\in Y} \{\langle x^*, x\rangle - g(y) - \delta_{G(\mathcal{C})}(x,y)\}.$$

Defining $F_1, F_2 : X \times Y \to \overline{\mathbb{R}}$ by $F_1(x,y) = g(y)$ and $F_2(x,y) = \delta_{G(\mathcal{C})}(x,y)$, it follows that for all $x^* \in X^*$ $\varphi^*(x^*) = (F_1 + F_2)^*(x^*, 0)$. The functions F_1 and F_2 are proper, convex and lower semicontinuous and the intersection of their domains is nonempty. Therefore, by Theorem 9.2, one has that for all $x^* \in X^*$

$$(F_1 + F_2)^*(x^*, 0) = \min_{(u^*,v^*)\in X^*\times Y^*} \{F_1^*(u^*, v^*) + F_2^*(x^* - u^*, -v^*)\} \ \forall x^* \in X^* \tag{9.5}$$

if and only if

epi F_1^* + epi F_2^* is closed regarding the set $X^* \times \{0\} \times \mathbb{R}$ in $(X^*, \omega(X^*, X)) \times (Y^*, \omega(Y^*, Y)) \times \mathbb{R}$.

For $(u^*, v^*) \in X^* \times Y^*$ we have

$$F_1^*(u^*, v^*) = \begin{cases} g^*(v^*), & \text{if } u^* = 0,\\ +\infty, & \text{otherwise,}\end{cases}$$

and

$$F_2^*(u^*, v^*) = \begin{cases} 0, & \text{if } (-u^*, -v^*) \in (G(\mathcal{C}))^*,\\ +\infty, & \text{otherwise}\end{cases}$$

and so, the equality in (9.5) is nothing else than

$$\varphi^*(x^*) = \min_{v^*\in Y^*} \{g^*(v^*) : (-x^*, v^*) \in (G(\mathcal{C}))^*\}$$

$$= \min_{v^*\in Y^*} \{g^*(v^*) : (v^*, x^*) \in G(\mathcal{C}^*)\} = \min_{v^*\in Y^*} \{g^*(v^*) : x^* \in \mathcal{C}^*(v^*)\} \ \forall x^* \in X^*.$$

This leads to the following statement.

Theorem 9.10. *We have*

$$\varphi^*(x^*) = \min_{v^*\in Y^*} \{g^*(v^*) : x^* \in \mathcal{C}^*(v^*)\} \ \forall x^* \in X^* \tag{9.6}$$

if and only if

$$\{0\} \times \operatorname{epi} g^* - (G(\mathcal{C}))^* \times \mathbb{R}_- \text{ is closed regarding the set } X^* \times \{0\} \times \mathbb{R} \text{ in } (X^*, \omega(X^*, X)) \times (Y^*, \omega(Y^*, Y)) \times \mathbb{R}. \tag{9.7}$$

Remark 9.11. Different sufficient conditions which ensure formula (9.6) have been given in [127, Theorem 2.8.6]. In case $A : X \to Y$ is a linear continuous operator and $\mathcal{C}$ is defined as $\mathcal{C}(x) = \{Ax\}$, $x \in X$, then φ turns out to be $g \circ A$, while the equivalence in Theorem 9.10 becomes

$$(g \circ A)^*(x^*) = \min_{y^* \in Y^*, x^* = A^* y^*} \{g^*(y^*)\} \ \forall x^* \in X^*$$

if and only if

$$(A^* \times \mathrm{id}_{\mathbb{R}})(\operatorname{epi} g^*) \text{ is closed in } (X^*, \omega(X^*, X)) \times \mathbb{R}.$$

In this way, we rediscover the statement of Theorem 7.11.

Chapter III
Biconjugate Functions

10 The Biconjugate of a General Perturbation Function

As follows from the Fenchel–Moreau Theorem, when the dual of a normed space X is endowed with the weak* topology, the biconjugate of a proper, convex and lower semicontinuous function defined on X coincides with the function itself. This is not necessarily the case when X^* is endowed with the strong topology. Working in the latter setting, we give in this chapter formulae for the biconjugates for some classes of functions, which appear in the convex optimization, that hold provided the validity of some suitable regularity conditions.

Consider X a normed space with the norm $\|\cdot\|$ and X^* its topological dual space, the norm of which being denoted by $\|\cdot\|_*$. On this space we work with three topologies, namely the strong one induced by $\|\cdot\|_*$ which attaches X^{**} as dual to X^*, the weak* one induced by X on X^*, $\omega(X^*, X)$, which makes X to be the dual of X^* and the weak one induced by X^{**} on X^*, $\omega(X^*, X^{**})$, that is the weakest topology on X^* which attaches X^{**} as dual to X^*. We specify each time when a weak topology is used, otherwise the strong one is considered. Like in the previous section, for sets and functions the closures and the lower semicontinuous hulls, respectively, in the weak* topologies are denoted by cl_{ω^*}, while the ones in the weak topologies are denoted by cl_{ω}. A normed space X can be identified with a subspace of X^{**}, and we denote by $\widehat{x}$ the canonical image in X^{**} of the element $x \in X$, i.e. $\langle \widehat{x}, x^* \rangle = \langle x^*, x \rangle$ for all $x \in X$ and $x^* \in X^*$, where by $\langle \cdot, \cdot \rangle$ we denote the duality product in both $X^* \times X$ and $X^{**} \times X^*$. For $U \subseteq X$ denote also $\widehat{U} = \{\widehat{x} : x \in U\}$. In this setting, when $f : X \to \overline{\mathbb{R}}$ is a given function, its biconjugate $f^{**} : X^{**} \to \overline{\mathbb{R}}$ is defined by $f^{**}(x^{**}) = \sup\{\langle x^{**}, x^* \rangle - f^*(x^*) : x^* \in X^*\}$.

For the beginning we prove the following preliminary result.

Lemma 10.1. *Let X be a normed space and let the convex function $f : X \to \overline{\mathbb{R}}$ have a nonempty domain. To f we attach the function*

$$\widehat{f} : X^{**} \to \overline{\mathbb{R}},\ \widehat{f}(x^{**}) = \begin{cases} f(x), & \text{if } x^{**} = \widehat{x}, \\ +\infty, & \text{otherwise}. \end{cases}$$

If $\mathrm{cl}_{\omega^*} \widehat{f}$ is proper, then $f^{**} = \mathrm{cl}_{\omega^*} \widehat{f}$.

R.I. Boţ, *Conjugate Duality in Convex Optimization*, Lecture Notes in Economics and Mathematical Systems 637, DOI 10.1007/978-3-642-04900-2_3,

Proof. Fix an $x^{**} \in X^{**}$. Then one has

$$f^{**}(x^{**}) = \sup_{x^* \in X^*} \{\langle x^{**}, x^* \rangle - f^*(x^*)\} = \sup_{\substack{x^* \in X^*, r \in \mathbb{R} \\ r \geq f^*(x^*)}} \{\langle x^{**}, x^* \rangle - r\}$$

$$= \sup_{\substack{x^* \in X^*, r \in \mathbb{R} \\ f(y) \geq \langle x^*, y \rangle - r \ \forall y \in X}} \{\langle x^{**}, x^* \rangle - r\} = \sup_{\substack{x^* \in X^*, r \in \mathbb{R} \\ \widehat{f}(y^{**}) \geq \langle y^{**}, x^* \rangle - r \ \forall y^{**} \in X}} \{\langle x^{**}, x^* \rangle - r\}$$

$$= \sup_{\substack{x^* \in X^*, r \in \mathbb{R} \\ \mathrm{cl}_{\omega^*} \widehat{f}(y^{**}) \geq \langle y^{**}, x^* \rangle - r \ \forall y^{**} \in X^{**}}} \{\langle x^{**}, x^* \rangle - r\} = \mathrm{cl}_{\omega^*} \widehat{f}(x^{**}).$$

Having the assumed properness and the obvious convexity of $\mathrm{cl}_{\omega^*} \widehat{f}$, which is also $\omega^*(X^{**}, X^*)$-lower semicontinuous, we use here that it coincides with the pointwise supremum of the set of its affine minorants (see [67]). □

The following corollary, which follows as a particular case of Lemma 10.1, was proved in [55, Section 4].

Corollary 10.2. *If X is a normed space, for a nonempty convex set $U \subseteq X$ one has $(\delta_U)^{**} = \delta_{\mathrm{cl}_{\omega^*}(\widehat{U})}$.*

Consider Y another normed space and let Y^* be its topological dual. Since there is no possibility of confusion, we denote the norms on X and Y by $\|\cdot\|$ and the ones on their duals by $\|\cdot\|_*$. We have the following general result (see [21]).

Theorem 10.3. *Let $\Phi : X \times Y \to \overline{\mathbb{R}}$ be a proper and convex function such that $0 \in \mathrm{Pr}_Y(\mathrm{dom}\, \Phi)$. Then $(\Phi(\cdot, 0))^{**} \geq \Phi^{**}(\cdot, 0)$. If Φ is also lower semicontinuous, then $(\Phi(\cdot, 0))^{**} = \Phi^{**}(\cdot, 0)$ if and only if $\mathrm{cl}_{\omega^*}\left(\inf_{y^* \in Y^*} \Phi^*(\cdot, y^*)\right) = \mathrm{cl}\left(\inf_{y^* \in Y^*} \Phi^*(\cdot, y^*)\right)$.*

Proof. The first inequality follows from $(\Phi(\cdot, 0))^* \leq \inf_{y^* \in Y^*} \Phi^*(\cdot, y^*)$, by considering the conjugates of these two functions. Take Φ moreover lower semicontinuous. Then, using Theorem 5.1, one gets

$$(\Phi(\cdot, 0))^* = \mathrm{cl}_{\omega^*}\left(\inf_{y^* \in Y^*} \Phi^*(\cdot, y^*)\right) \leq \mathrm{cl}_{\omega}\left(\inf_{y^* \in Y^*} \Phi^*(\cdot, y^*)\right)$$

$$= \mathrm{cl}\left(\inf_{y^* \in Y^*} \Phi^*(\cdot, y^*)\right) \leq \inf_{y^* \in Y^*} \Phi^*(\cdot, y^*),$$

from which, by considering the conjugates and taking into consideration that a function and its closure have the same conjugate, follows

$$(\Phi(\cdot, 0))^{**} = \left(\mathrm{cl}_{\omega^*}\left(\inf_{y^* \in Y^*} \Phi^*(\cdot, y^*)\right)\right)^* \geq \left(\mathrm{cl}\left(\inf_{y^* \in Y^*} \Phi^*(\cdot, y^*)\right)\right)^*$$

$$= \left(\inf_{y^* \in Y^*} \Phi^*(\cdot, y^*) \right)^* = \Phi^{**}(\cdot, 0).$$

It is straightforward that if the closure of $\inf_{y^* \in Y^*} \Phi^*(\cdot, y^*)$ coincides with its weak* closure one gets $(\Phi(\cdot, 0))^{**} = \Phi^{**}(\cdot, 0)$. Assume now this equality true. Then by the previous relation we have

$$\left(\mathrm{cl}_{\omega^*} \left(\inf_{y^* \in Y^*} \Phi^*(\cdot, y^*) \right) \right)^* = \left(\mathrm{cl} \left(\inf_{y^* \in Y^*} \Phi^*(\cdot, y^*) \right) \right)^*,$$

followed by

$$\left(\mathrm{cl}_{\omega^*} \left(\inf_{y^* \in Y^*} \Phi^*(\cdot, y^*) \right) \right)^{**} (x^{***}) = \left(\mathrm{cl} \left(\inf_{y^* \in Y^*} \Phi^*(\cdot, y^*) \right) \right)^{**} (x^{***}) \ \forall x^{***} \in X^{***}. \tag{10.1}$$

The weak* closure of $\inf_{y^* \in Y^*} \Phi^*(\cdot, y^*)$ is proper and convex (see, for instance, the proof of Theorem 5.1) and also weak* lower semicontinuous, which yields that it is lower semicontinuous, too. On the other hand, $\mathrm{cl}\left(\inf_{y^* \in Y^*} \Phi^*(\cdot, y^*)\right)$ is convex and lower semicontinuous and also proper because it is greater than or equal to the proper function $\mathrm{cl}_{\omega^*}\left(\inf_{y^* \in Y^*} \Phi^*(\cdot, y^*)\right)$ and if it would be everywhere equal to $+\infty$ then Φ^* would be everywhere $+\infty$, too. But when this happens, then for all $x \in X$ $\Phi(x, 0) = \Phi^{**}(\widehat{x}, 0) = -\infty$, which contradicts the properness of Φ. By the Fenchel–Moreau Theorem we obtain for all $x^* \in X^*$ that $\left(\mathrm{cl}_{\omega^*}\left(\inf_{y^* \in Y^*} \Phi^*(\cdot, y^*)\right)\right)^{**}(\widehat{x^*}) = \mathrm{cl}_{\omega^*}\left(\inf_{y^* \in Y^*} \Phi^*(\cdot, y^*)\right)(x^*)$ and $\left(\mathrm{cl}\left(\inf_{y^* \in Y^*} \Phi^*(\cdot, y^*)\right)\right)^{**}(\widehat{x^*}) = \mathrm{cl}\left(\inf_{y^* \in Y^*} \Phi^*(\cdot, y^*)\right)(x^*)$ $\forall x^* \in X^*$. Finally, (10.1) yields $\mathrm{cl}_{\omega^*}\left(\inf_{y^* \in Y^*} \Phi^*(\cdot, y^*)\right) = \mathrm{cl}\left(\inf_{y^* \in Y^*} \Phi^*(\cdot, y^*)\right)$. □

Theorem 10.4. *Let $\Phi : X \times Y \to \overline{\mathbb{R}}$ be a proper, convex and lower semicontinuous function such that $0 \in \mathrm{Pr}_Y(\mathrm{dom}\,\Phi)$. If one of the following conditions:*

(i) $\mathrm{Pr}_{X^* \times \mathbb{R}}(\mathrm{epi}\,\Phi^*)$ is closed in $(X^*, \omega(X^*, X)) \times \mathbb{R}$;
(ii) X and Y are Banach spaces and $0 \in \mathrm{sqri}(\mathrm{Pr}_Y(\mathrm{dom}\,\Phi))$;

is satisfied, then $(\Phi(\cdot, 0))^{**} = \Phi^{**}(\cdot, 0)$.

Proof. By Theorem 5.5 it follows that (ii) implies (i), which is nothing else than (cf. Corollary 5.3) $(\Phi(\cdot, 0))^*(x^*) = \min_{y^* \in Y^*} \Phi^*(x^*, y^*)$ for all $x^* \in X^*$. From (5.1) it follows that

$$\mathrm{cl}_{\omega^*} \left(\inf_{y^* \in Y^*} \Phi^*(\cdot, y^*) \right) = \mathrm{cl} \left(\inf_{y^* \in Y^*} \Phi^*(\cdot, y^*) \right) = \inf_{y^* \in Y^*} \Phi^*(\cdot, y^*)$$

and Theorem 10.3 leads to the desired conclusion. □

Remark 10.5. One should notice that for proving relation $(\Phi(\cdot, 0))^{**} \geq \Phi^{**}(\cdot, 0)$ in Theorem 10.3, we use neither the properness nor the convexity assumed for Φ. On the other hand, since the epigraph of the closure of the infimal value function of Φ^* coincides with the closure of the projection of epi Φ^* on $X^* \times \mathbb{R}$, the equalities in Theorem 10.3 are further equivalent to

$$\mathrm{cl}_{\omega^*\times\mathcal{R}}\left(\mathrm{Pr}_{X^*\times\mathbb{R}}(\mathrm{epi}(\Phi^*)\right) = \mathrm{cl}\left(\mathrm{Pr}_{X^*\times\mathbb{R}}(\mathrm{epi}(\Phi^*))\right).$$

Notice also that, renouncing in Theorem 10.4 the lower semicontinuity assumption for Φ, but asking instead that the regularity conditions (RC_1^Φ) is fulfilled, one gets the same conclusion.

11 Biconjugates Formulae for Different Classes of Convex Functions

We give in this section formulae for the biconjugate functions of some functions that appear often in convex optimization by deducing them from the general results provided in Section 10. The convex functions we investigate here are the ones for which we have given in Sections 6–8 generalized Moreau–Rockafellar results.

We start by considering the proper and convex functions $f : X \to \overline{\mathbb{R}}$ and $g : Y \to \overline{\mathbb{R}}$ and the linear continuous operator $A : X \to Y$ fulfilling the feasibility condition $A(\mathrm{dom}\, f) \cap \mathrm{dom}\, g \neq \emptyset$. We are interested in giving a formula for the biconjugate of the function $f + g \circ A$. In order to do this, we take again as perturbation function $\Phi^A : X \times Y \to \overline{\mathbb{R}}$, $\Phi^A(x, y) = f(x) + g(Ax + y)$. By (2.1), the conjugate of Φ^A is $(\Phi^A)^* : X^* \times Y^* \to \overline{\mathbb{R}}$, $(\Phi^A)^*(x^*, y^*) = f^*(x^* - A^*y^*) + g^*(y^*)$. After a simple calculation, its biconjugate turns out to be

$$(\Phi^A)^{**}(x^{**}, y^{**}) : X^{**} \times Y^{**} \to \overline{\mathbb{R}},\ (\Phi^A)^{**}(x^{**}, y^{**})$$
$$= f^{**}(x^{**}) + g^{**}(A^{**}x^{**} + y^{**}).$$

It holds $\mathrm{Pr}_Y(\mathrm{dom}\,\Phi^A) = \mathrm{dom}\, g - A(\mathrm{dom}\, f)$ (cf. Section 2), $\inf_{y^*\in Y^*}(\Phi^A)^*(\cdot, y^*) = f^*\Box A^*g^*$ and $\mathrm{Pr}_{X^*\times\mathbb{R}}(\mathrm{epi}(\Phi^A)^*) = \mathrm{epi}\, f^* + (A^* \times \mathrm{id}_{\mathbb{R}})(\mathrm{epi}\, g^*)$ (cf. Section 7) and so, from Theorems 10.3 and 10.4, we get the following statement.

Theorem 11.1. *One always has* $(f + g \circ A)^{**} \geq f^{**} + g^{**} \circ A^{**}$. *Assuming the additional hypothesis*

(H^A) f *and* g *are lower semicontinuous*

fulfilled, it follows that the inequality above is always fulfilled as equality if and only if $\mathrm{cl}_{\omega^*}(f^*\Box A^*g^*) = \mathrm{cl}(f^*\Box A^*g^*)$. *Under* (H^A), *if one of the following conditions:*

(i) $\mathrm{epi}\, f^* + (A^* \times \mathrm{id}_{\mathbb{R}})(\mathrm{epi}\, g^*)$ *is closed in* $(X^*, \omega(X^*, X)) \times \mathbb{R}$;
(ii) X *and* Y *are Banach spaces and* $0 \in \mathrm{sqri}(\mathrm{dom}\, g - A(\mathrm{dom}\, f))$;

is satisfied, then $(f + g \circ A)^{**} = f^{**} + g^{**} \circ A^{**}$.

When taking $X = Y$ and $A = \mathrm{id}_X$ we obtain the following statement as a particular case of the previous theorem.

Theorem 11.2. *One always has $(f+g)^{**} \geq f^{**}+g^{**}$. Assuming the additional hypothesis*

$$(H^{\mathrm{id}})\ f \text{ and } g \text{ are lower semicontinuous}$$

fulfilled, it follows that the inequality above is always fulfilled as equality if and only if $\mathrm{cl}_{\omega^*}(f^*\square g^*) = \mathrm{cl}(f^*\square g^*)$. Under (H^{id}), if one of the following conditions:

(i) $\operatorname{epi} f^* + \operatorname{epi} g^*$ *is closed in* $(X^*, \omega(X^*, X)) \times \mathbb{R}$;
(ii) X *is a Banach space and* $0 \in \operatorname{sqri}(\operatorname{dom} g - \operatorname{dom} f)$;

is satisfied, then $(f+g)^{**} = f^{**}+g^{**}$.

Assuming in Theorem 11.1 that f is identical to 0 we obtain the following assertion.

Theorem 11.3. *One always has $(g \circ A)^{**} \geq g^{**} \circ A^{**}$. Assuming the additional hypothesis*

$$(H^{A_g})\ g \text{ is lower semicontinuous}$$

fulfilled, it follows that the inequality above is always fulfilled as equality if and only if $\mathrm{cl}_{\omega^*}(A^*g^*) = \mathrm{cl}(A^*g^*)$. Under (H^{A_g}), if one of the following conditions:

(i) $(A^* \times \mathrm{id}_{\mathbb{R}})(\operatorname{epi} g^*)$ *is closed in* $(X^*, \omega(X^*, X)) \times \mathbb{R}$;
(ii) X *and* Y *are Banach spaces and* $0 \in \operatorname{sqri}(\operatorname{dom} g - R(A))$;

is satisfied, then $(g \circ A)^{**} = g^{**} \circ A^{**}$.

For $m \geq 2$ considering $f_i : X \to \overline{\mathbb{R}}$, $i = 1, \ldots m$, proper and convex functions fulfilling $\cap_{i=1}^m \operatorname{dom} f_i \neq \emptyset$, $Y = X^m$, $g : X^m \to \overline{\mathbb{R}}$, $g(x^1, \ldots, x^m) = \sum_{i=1}^m f_i(x^i)$ and $A : X \to X^m$, $Ax = (x, \ldots, x)$, one has (cf. Sections 2 and 7) $\operatorname{dom} g - R(A) = \prod_{i=1}^m \operatorname{dom} f_i - \Delta_{X^m}$, $A^*g^* = f_1^*\square \ldots \square f_m^*$ and $(A^* \times \mathrm{id}_{\mathbb{R}})(\operatorname{epi} g^*) = \sum_{i=1}^m \operatorname{epi} f_i^*$. Moreover, $g^{**}((x^1)^{**}, \ldots, (x^m)^{**}) = \sum_{i=1}^m f_i^{**}((x^i)^{**})$ and $A^{**}(x^{**}) = (x^{**}, \ldots, x^{**})$. Theorem 11.3 leads to the following result.

Theorem 11.4. *One always has $(\sum_{i=1}^m f_i)^{**} \geq \sum_{i=1}^m f_i^{**}$. Assuming the additional hypothesis*

$$(H^{\Sigma})\ f_i \text{ is lower semicontinuous, } i = 1, \ldots, m,$$

fulfilled, it follows that the inequality above is always fulfilled as equality if and only if $\mathrm{cl}_{\omega^*}(f_1^*\square \ldots \square f_m^*) = \mathrm{cl}(f_1^*\square \ldots \square f_m^*)$. Under (H^{Σ}), if one of the following conditions:

(i) $\sum_{i=1}^m \operatorname{epi} f_i^*$ *is closed in* $(X^*, \omega(X^*, X)) \times \mathbb{R}$;
(ii) X *is a Banach space and* $0 \in \operatorname{sqri}\left(\prod_{i=1}^m \operatorname{dom} f_i - \Delta_{X^m}\right)$;

is satisfied, then $(\sum_{i=1}^m f_i)^{**} = \sum_{i=1}^m f_i^{**}$.

One can notice that in the last three theorems we rediscover and partially extend some results given in [129]. The statement proved in Theorem 11.2(ii), namely that when X is a Banach space, $f, g : X \to \overline{\mathbb{R}}$ are proper, convex and lower semicontinuous functions fulfilling $0 \in \operatorname{sqri}(\operatorname{dom} g - \operatorname{dom} f)$, one has $(f+g)^{**} = f^{**}+g^{**}$ is well-known in the literature (see [71]). Rockafellar has used in [114, Proposition 1] the equality $(f + g)^{**} = f^{**} + g^{**}$, when proving that the subdifferential of a proper, convex and lower semicontinuous function on X is maximal monotone.

The setting we consider next is the following: X and Z are separated locally convex spaces, where Z is partially ordered by the nonempty convex cone $C \subseteq Z$, $S \subseteq X$ is a nonempty and convex set, $f : X \to \overline{\mathbb{R}}$ a proper and convex function and $g : X \to Z^\bullet$ a proper and C-convex function such that $\operatorname{dom} f \cap S \cap g^{-1}(-C) \neq \emptyset$. We take for Φ the three perturbation functions which provided the Lagrange, Fenchel and Fenchel–Lagrange dual problems to (P^C). We start with $\Phi^{C_L} : X \times Z \to \overline{\mathbb{R}}$, $\Phi^{C_L}(x, z) = f(x) + \delta_{\{u \in S : g(u) \in z - C\}}(x)$. By (3.1), its conjugate is $(\Phi^{C_L})^* : X^* \times Z^* \to \overline{\mathbb{R}}$,

$$(\Phi^{C_L})^*(x^*, z^*) = \begin{cases} (f + ((-z^*)g) + \delta_S)^*(x^*), & \text{if } z^* \in -C^*, \\ +\infty, & \text{otherwise.} \end{cases}$$

Further, one can determine the second conjugate of Φ^{C_L}, which is $(\Phi^{C_L})^{**} : X^{**} \times Z^{**} \to \overline{\mathbb{R}}$,

$$(\Phi^{C_L})^{**}(x^{**}, z^{**}) = \sup_{z^* \in C^*} \{-\langle z^{**}, z^* \rangle + (f + (z^* g) + \delta_S)^{**}(x^{**})\}.$$

We found out that $\operatorname{Pr}_Z(\operatorname{dom} \Phi^{C_L}) = g(\operatorname{dom} f \cap S \operatorname{dom} g) + C$ (cf. Section 3), $\inf_{z^* \in Z^*} (\Phi^{C_L})^*(\cdot, z^*) = \inf_{z^* \in C^*} (f + (z^* g) + \delta_S)^*$ and $\operatorname{Pr}_{X^* \times \mathbb{R}}(\operatorname{epi}(\Phi^{C_L})^*) = \cup_{z^* \in C^*} \operatorname{epi}(f + (z^* g) + \delta_S)^*$ (cf. Section 8) and Theorem 10.3 and, consequently, Theorem 10.4 furnish the following result.

Theorem 11.5. *One always has* $(f + \delta_{\mathcal{A}})^{**} \geq \sup_{z^* \in C^*} (f + (z^* g) + \delta_S)^{**}$. *Assuming the additional hypothesis*

(H^{C_L}) *S is closed, f is lower semicontinuous and g is C-epi closed*

fulfilled, it follows that the inequality above is always fulfilled as equality if and only if $\operatorname{cl}_{\omega^*}(\inf_{z^* \in C^*} (f + (z^* g) + \delta_S)^*) = \operatorname{cl}(\inf_{z^* \in C^*} (f + (z^* g) + \delta_S)^*)$. *Under* (H^{C_L}), *if one of the following conditions:*

(i) $\bigcup_{z^* \in C^*} \operatorname{epi}(f + (z^* g) + \delta_S)^*$ *is closed in* $(X^*, \omega(X^*, X)) \times \mathbb{R}$;
(ii) X and Z are Banach spaces and $0 \in \operatorname{sqri}(g(\operatorname{dom} f \cap S \cap \operatorname{dom} g) + C)$;

is satisfied, then $(f + \delta_{\mathcal{A}})^{**} = \sup_{z^* \in C^*} (f + (z^* g) + \delta_S)^{**}$.

Considering $\Phi^{C_F} : X \times X \to \overline{\mathbb{R}}$, $\Phi^{C_F}(x, y) = f(x+y) + \delta_{\mathcal{A}}(x)$, as perturbation function, one can employ Corollary 10.2 and Theorem 11.2 for obtaining another formula for the biconjugate of $f + \delta_{\mathcal{A}}$.

Theorem 11.6. *One always has* $(f + \delta_{\mathcal{A}})^{**} \geq f^{**} + \delta_{\mathrm{cl}_{\omega^*}(\widehat{\mathcal{A}})}$. *Assuming the additional hypothesis*

$$(H^{C_F})\ f \text{ is lower semicontinuous and } \mathcal{A} \text{ is closed}$$

fulfilled, it follows that the inequality above is always fulfilled as equality if and only if $\mathrm{cl}_{\omega^*}(f^*\Box\sigma_{\mathcal{A}}) = \mathrm{cl}(f^*\Box\sigma_{\mathcal{A}})$. *Under* (H^{C_F}), *if one of the following conditions:*

(i) $\mathrm{epi}\, f^* + \mathrm{epi}\, \sigma_{\mathcal{A}}$ *is closed in* $(X^*, \omega(X^*, X)) \times \mathbb{R}$;
(ii) X *is a Banach space and* $0 \in \mathrm{sqri}(\mathrm{dom}\, f - \mathcal{A})$;

is satisfied, then $(f + \delta_{\mathcal{A}})^{**} = f^{**} + \delta_{\mathrm{cl}_{\omega^*}(\widehat{\mathcal{A}})}$.

If we want to separate f from g in the formula of $(f + \delta_{\mathcal{A}})^{**}$, a good choice is to consider the perturbation function $\Phi^{C_{FL}} : X \times X \times Z \to \overline{\mathbb{R}}$, $\Phi^{C_{FL}}(x, y, z) = f(x + y) + \delta_{\{u \in S : g(u) \in z - C\}}(x)$, which leads to the Fenchel–Lagrange dual problem to (P^C). Its conjugate $(\Phi^{C_{FL}})^* : X^* \times X^* \times Z^* \to \overline{\mathbb{R}}$ is (cf. (3.3)),

$$(\Phi^{C_{FL}})^*(x^*, y^*, z^*) = \begin{cases} f^*(y^*) + (((-z^*)g) + \delta_S)^*(x^* - y^*), & \text{if } z^* \in -C^*, \\ +\infty, & \text{otherwise,} \end{cases}$$

while its second conjugate $(\Phi^{C_{FL}})^{**} : X^{**} \times X^{**} \times Z^{**} \to \overline{\mathbb{R}}$ turns out to be

$$(\Phi^{C_{FL}})^{**}(x^{**}, y^{**}, z^{**}) = f^{**}(x^{**} + y^{**}) + \sup_{z^* \in C^*} \{-\langle z^{**}, z^* \rangle + ((z^*g) + \delta_S)^{**}(x^{**})\}.$$

For the next theorem, we take into consideration the fact that $\mathrm{Pr}_{Y \times Z}(\mathrm{dom}\, \Phi^{C_{FL}}) = \mathrm{dom}\, f \times C - \mathrm{epi}_{(-C)}(-g) \cap (S \times Z)$ (cf. Section 3), $\inf_{y^* \in X^*, z^* \in Z^*} (\Phi^{C_{FL}})^*(\cdot, y^*, z^*) = \inf_{z^* \in C^*} f^*\Box((z^*g) + \delta_S)^*$ and $\mathrm{Pr}_{X^* \times \mathbb{R}}(\mathrm{epi}(\Phi^{C_{FL}})^*) = \mathrm{epi}\, f^* + \cup_{z^* \in C^*} \mathrm{epi}((z^*g) + \delta_S)^*$ (cf. Section 8) and obtain via Theorems 10.3 and 10.4 the following result.

Theorem 11.7. *One always has* $(f + \delta_{\mathcal{A}})^{**} \geq f^{**} + \sup_{z^* \in C^*}((z^*g) + \delta_S)^{**}$. *Assuming the additional hypothesis* (H^{C_L}) *fulfilled, it follows that the inequality above is always fulfilled as equality if and only if* $\mathrm{cl}_{\omega^*}(\inf_{z^* \in C^*} f^*\Box((z^*g) + \delta_S)^*) = \mathrm{cl}(\inf_{z^* \in C^*} f^*\Box((z^*g) + \delta_S)^*)$. *Under* (H^{C_L}), *if one of the following conditions:*

(i) $\mathrm{epi}\, f^* + \bigcup_{z^* \in C^*} \mathrm{epi}((z^*g) + \delta_S)^*$ *is closed in* $(X^*, \omega(X^*, X)) \times \mathbb{R}$;
(ii) X *and* Z *are Banach spaces and*
$0 \in \mathrm{sqri}\big(\mathrm{dom}\, f \times C - \mathrm{epi}_{(-C)}(-g) \cap (S \times Z)\big)$;

is satisfied, then $(f + \delta_{\mathcal{A}})^{**} = f^{**} + \sup_{z^* \in C^*}((z^*g) + \delta_S)^{**}$.

We close the section by giving formulae for the biconjugate of the composed convex function $f + g \circ h$, by working in the same framework as in Section 4, more precisely by considering X and Z be separated locally convex spaces, where Z is partially ordered by the nonempty convex cone $C \subseteq Z$, $f : X \to \overline{\mathbb{R}}$ a proper and convex function, $g : Z \to \overline{\mathbb{R}}$ a proper, convex and C-increasing function fulfilling, by convention, $g(\infty_C) = +\infty$ and $h : X \to Z^\bullet = Z \cup \{\infty_C\}$ a proper and C-convex function such that $h(\operatorname{dom} f \cap \operatorname{dom} h) \cap \operatorname{dom} g \neq \emptyset$.

We take first as perturbation function $\Phi^{CC_1} : X \times Z \to \overline{\mathbb{R}}$, $\Phi^{CC_1}(x,z) = f(x) + g(h(x)+z)$. Its conjugate is (cf. (4.1)) $(\Phi^{CC_1})^* : X^* \times Z^* \to \overline{\mathbb{R}}$,

$$(\Phi^{CC_1})^*(x^*,z^*) = \begin{cases} g^*(z^*) + (f + (z^*h))^*(x^*), & \text{if } z^* \in C^*, \\ +\infty, & \text{otherwise,} \end{cases}$$

and from here one obtains for the biconjugate of Φ^{CC_1}, $(\Phi^{CC_1})^{**} : X^{**} \times Z^{**} \to \overline{\mathbb{R}}$, the following expression

$$(\Phi^{CC_1})^{**}(x^{**},z^{**}) = \sup_{\substack{x^* \in X^* \\ z^* \in \operatorname{dom} g^*}} \{\langle x^{**},x^*\rangle + \langle z^{**},z^*\rangle - g^*(z^*) - (f+(z^*h))^*(x^*)\}$$

$$= \sup_{z^* \in \operatorname{dom} g^*} \{\langle z^{**},z^*\rangle - g^*(z^*) + (f + (z^*h))^{**}(x^{**})\}.$$

As we have seen in the previous section, $\operatorname{Pr}_Z(\operatorname{dom}\Phi^{CC_1}) = \operatorname{dom} g - h(\operatorname{dom} f \cap \operatorname{dom} h)$ (cf. Section 4), $\inf_{z^* \in Z^*}(\Phi^{CC_1})^*(\cdot,z^*) = \inf_{z^* \in C^*}\{g^*(z^*)+(f+(z^*h))^*(\cdot)\}$ and $\operatorname{Pr}_{X^* \times \mathbb{R}}(\operatorname{epi}(\Phi^{CC_1})^*) = \cup_{z^* \in \operatorname{dom} g^*}((0,g^*(z^*)) + \operatorname{epi}(f + (z^*h))^*$ (cf. Section 6). Moreover, in case f and g are lower semicontinuous and h is star C-lower semicontinuous, the perturbation function Φ^{CC_1} turns out to be lower semicontinuous, too (see Section 4). Consequently, by Theorems 10.3 and 10.4, we obtain the following statement concerning $(f + g \circ h)^{**}$.

Theorem 11.8. *One always has*

$$(f + g \circ h)^{**} \geq \sup_{z^* \in \operatorname{dom} g^*} \{-g^*(z^*) + (f + (z^*h))^{**}(\cdot)\}.$$

Assuming the additional hypothesis
(H^{CC}) f and g are lower semicontinuous and h is star C-lower semicontinuous fulfilled, it follows that the inequality above is always fulfilled as equality if and only if

$$\operatorname{cl}_{\omega^*}\left(\inf_{z^* \in C^*}\{g^*(z^*)+(f+(z^*h))^*(\cdot)\}\right) = \operatorname{cl}\left(\inf_{z^* \in C^*}\{g^*(z^*)+(f+(z^*h))^*(\cdot)\}\right).$$

Under (H^{CC}), if one of the following conditions:

(i) $\cup_{z^* \in \operatorname{dom} g^*}((0,g^*(z^*)) + \operatorname{epi}(f + (z^*h))^*)$ *is closed in* $(X^*, \omega(X^*,X)) \times \mathbb{R}$*;*
(ii) X *and* Z *are Banach spaces and* $0 \in \operatorname{sqri}(\operatorname{dom} g - h(\operatorname{dom} f \cap \operatorname{dom} h))$*;*

is satisfied, then $(f + g \circ h)^{**} = \sup_{z^* \in \operatorname{dom} g^*}\{-g^*(z^*) + (f + (z^*h))^{**}(\cdot)\}$.

If we want to have f separated from h in the formula of the biconjugate, the perturbation function $\Phi^{CC_2} : X \times X \times Z \to \overline{\mathbb{R}}$, $\Phi^{CC_2}(x, y, z) = f(x + y) + g(h(x) + z)$, can be considered. Its conjugate is (cf. (4.2)) $(\Phi^{CC_1})^* : X^* \times X^* \times Z^* \to \overline{\mathbb{R}}$

$$(\Phi^{CC_2})^*(x^*, y^*, z^*) = \begin{cases} g^*(z^*) + f^*(y^*) + (z^*h)^*(x^* - y^*), & \text{if } z^* \in C^*, \\ +\infty, & \text{otherwise,} \end{cases}$$

while its biconjugate $(\Phi^{CC_2})^{**} : X^{**} \times X^{**} \times Z^{**} \to \overline{\mathbb{R}}$ turns out to be

$$(\Phi^{CC_2})^{**}(x^{**}, y^{**}, z^{**}) = \sup_{\substack{x^*, y^* \in X^* \\ z^* \in \operatorname{dom} g^*}} \{\langle x^{**}, x^*\rangle + \langle y^{**}, y^*\rangle + \langle z^{**}, z^*\rangle - g^*(z^*) - f^*(y^*)$$

$$-(z^*h)^*(x^* - y^*)\} = f^{**}(x^{**} + y^{**}) + \sup_{z^* \in \operatorname{dom} g^*} \{\langle z^{**}, z^*\rangle - g^*(z^*) + (z^*h)^{**}(x^{**})\}.$$

In this case, we have $\Pr_{X \times Z}(\operatorname{dom} \Phi^{CC_2}) = \operatorname{dom} f \times \operatorname{dom} g - \operatorname{epi}_C h$ (cf. Section 4), $\inf_{y^* \in X^*, z^* \in Z^*} (\Phi^{CC_2})^*(\cdot, y^*, z^*) = \inf_{z^* \in C^*} \{g^*(z^*) + f^* \square (z^*h)^*(\cdot)\}$ and $\Pr_{X^* \times \mathbb{R}}(\operatorname{epi}(\Phi^{CC_2})^*) = \operatorname{epi} f^* + \cup_{z^* \in \operatorname{dom} g^*} ((0, g^*(z^*)) + \operatorname{epi}(z^*h)^*$ (cf. Section 6). Noticing also that under (H^{CC}) Φ^{CC_2} is lower semicontinuous we obtain the following result.

Theorem 11.9. *One always has*

$$(f + g \circ h)^{**} \geq f^{**} + \sup_{z^* \in \operatorname{dom} g^*} \{-g^*(z^*) + (z^*h)^{**}(\cdot)\}.$$

Assuming the additional hypothesis (H^{CC}) fulfilled, it follows that the inequality above is always fulfilled as equality if and only if

$$\operatorname{cl}_{\omega^*} \left(\inf_{z^* \in C^*} \{g^*(z^*) + f^* \square (z^*h)^*(\cdot)\} \right) = \operatorname{cl} \left(\inf_{z^* \in C^*} \{g^*(z^*) + f^* \square (z^*h)^*(\cdot)\} \right).$$

Under (H^{CC}), if one of the following conditions:

(i) $\operatorname{epi} f^* + \cup_{z^* \in \operatorname{dom} g^*} ((0, g^*(z^*)) + \operatorname{epi}(z^*h)^*)$ *is closed in* $(X^*, \omega(X^*, X)) \times \mathbb{R}$*;*
(ii) X and Z are Banach spaces and $0 \in \operatorname{sqri}(\operatorname{dom} f \times \operatorname{dom} g - \operatorname{epi}_C h)$*;*

is satisfied, then $(f + g \circ h)^{**} = f^{**} + \sup_{z^* \in \operatorname{dom} g^*} \{-g^*(z^*) + (z^*h)^{**}(\cdot)\}$.

12 The Supremum of an (Infinite) Family of Convex Functions

In this section, we perform some investigations concerning the supremum of an (infinite) family of convex functions and provide several formulae and characterizations for its conjugate and biconjugate function. In the case of the latter we relax the hypothesis imposed on the space in which we work in the sense that instead of a normed space, as in the previous sections, it is taken to be only separated locally convex.

Consider a dual system formed by a separated locally convex vector space X and its topological dual space X^*, which is assumed for the beginning to be endowed with the weak* topology. Let the proper convex functions $f_t : X \to \overline{\mathbb{R}}$, $t \in T$, where T is an arbitrary index set, possibly uncountable, such that $\operatorname{dom}(\sup_{t\in T} f_t) \neq \emptyset$. This yields that the set $\cap_{t\in T} \operatorname{dom} f_t$ is nonempty. By $\mathbb{R}^T$ we denote the space of all functions $z : T \to \mathbb{R}$, endowed with the product topology and with the operations being the usual pointwise ones. For simplicity, we denote $z_t = z(t)$ for $z \in \mathbb{R}^T$ and all $t \in T$. Let $\Delta_{\mathbb{R}^T}$ be the subset of the constant functions $z \in \mathbb{R}^T$. The dual space of $\mathbb{R}^T$ is $(\mathbb{R}^T)^*$, the so-called *space of generalized finite sequences* $z^* = (z_t^*)_{t\in T}$ such that $z_t^* \in \mathbb{R}$ for all $t \in T$, and with only finitely many z_t^* different from zero. The positive cone in $\mathbb{R}^T$ is $\mathbb{R}^T_+ = \{z \in \mathbb{R}^T : z_t = z(t) \geq 0 \ \forall t \in T\}$, and its dual is the positive cone in $(\mathbb{R}^T)^*$, namely $(\mathbb{R}^T_+)^* = \{z^* = (z_t^*)_{t\in T} \in (\mathbb{R}^T)^* : z_t^* \geq 0 \ \forall t \in T\}$. Denote also $\mathcal{P} = \{z^* \in (\mathbb{R}^T_+)^* : \sum_{t\in T} z_t^* = 1\}$.

Further, we additionally assume that $f_t : X \to \overline{\mathbb{R}}$, $t \in T$, are lower semicontinuous and define

$$f : X \to (\mathbb{R}^T)^\bullet = \mathbb{R}^T \cup \{\infty_{\mathbb{R}^T_+}\},\ f(x) = \begin{cases} (f_t(x))_{t\in T}, & \text{if } x \in \cap_{t\in T} \operatorname{dom} f_t, \\ \infty_{\mathbb{R}^T_+}, & \text{otherwise.} \end{cases}$$

Evidently, f is proper, $\mathbb{R}^T_+$-convex and $\mathbb{R}^T_+$-epi closed.

One can easily notice that, when denoting by $|S|$ the *cardinality* of the set S, for all $x \in X$ there is

$$\sup_{t\in T} f_t(x) = \sup_{\substack{S\subseteq T \\ |S|<+\infty}} \sup_{\substack{z_s^*>0\ \forall s\in S \\ \sum_{s\in S} z_s^*=1}} \sum_{s\in S} z_s^* f_s(x).$$

For all the finite subsets S of T and any $z_s^* > 0, s \in S$, fulfilling $\sum_{s\in S} z_s^* = 1$, the function $x \mapsto \sum_{s\in S} z_s^* f_s(x)$ is proper, convex and lower semicontinuous. From the Fenchel–Moreau Theorem it follows that this function is equal to its biconjugate, thus for all $x \in X$ one has

$$\sup_{t\in T} f_t(x) = \sup_{\substack{S\subseteq T \\ |S|<+\infty}} \sup_{\substack{z_s^*>0\ \forall s\in S \\ \sum_{s\in S} z_s^*=1}} \left(\sum_{s\in S} z_s^* f_s\right)^{**}(x)$$

$$= \left(\inf_{\substack{S\subseteq T, |S|<+\infty \\ z_s^*>0\ \forall s\in S \\ \sum_{s\in S} z_s^*=1}} \left(\sum_{s\in S} z_s^* f_s\right)^* \right)^*(x). \tag{12.1}$$

Proposition 12.1. *The function*

$$\eta : X^* \to \overline{\mathbb{R}},\ \eta(x^*) = \inf_{\substack{S \subseteq T, |S| < +\infty \\ z_s^* > 0\ \forall s \in S \\ \sum_{s \in S} z_s^* = 1}} \left(\sum_{s \in S} z_s^* f_s \right)^* (x^*)$$

is proper and convex.

Proof. Whenever $S \subseteq T$ with S finite and $z_s^* > 0$ for all $s \in S$ with $\sum_{s \in S} z_s^* = 1$, the function $\left(\sum_{s \in S} z_s^* f_s\right)^*$ is proper, thus η cannot be identical to $+\infty$. Assuming the existence of some $x^* \in X^*$ where $\eta(x^*) = -\infty$, it follows that $\sup_{t \in T} f_t$ is identical to $+\infty$, which contradicts the feasibility hypothesis $\operatorname{dom}(\sup_{t \in T} f_t) \neq \emptyset$. Therefore, η is proper.

In order to prove its convexity, let some arbitrary $\lambda \in [0, 1]$ and $x_1^*, x_2^* \in X^*$. What we have to show is

$$\eta(\lambda x_1^* + (1 - \lambda) x_2^*) \leq \lambda \eta(x_1^*) + (1 - \lambda) \eta(x_2^*).$$

If $\lambda \in \{0, 1\}$ or $\eta(x_1^*) = +\infty$ or $\eta(x_2^*) = +\infty$, the inequality above is valid. Let further $\lambda \in (0, 1)$ and $\eta(x_1^*), \eta(x_2^*) \in \mathbb{R}$. Then, there are some $\alpha, \beta \in \mathbb{R}$ such that $\eta(x_1^*) < \alpha$ and $\eta(x_2^*) < \beta$. Thus, there exist the finite subsets S_1 and S_2 of T and $z_s^* > 0$ for all $s \in S_1 \cup S_2$ such that $\sum_{s \in S_1} z_s^* = 1$, $\sum_{s \in S_2} z_s^* = 1$, $\left(\sum_{s \in S_1} z_s^* f_s\right)^* (x_1^*) < \alpha$ and $\left(\sum_{s \in S_2} z_s^* f_s\right)^* (x_2^*) < \beta$. Consequently,

$$\eta(\lambda x_1^* + (1 - \lambda) x_2^*) \leq \left(\lambda \sum_{s \in S_1} z_s^* f_s + (1 - \lambda) \sum_{s \in S_2} z_s^* f_s \right)^* (\lambda x_1^* + (1 - \lambda) x_2^*)$$

$$\leq \left(\sum_{s \in S_1} \lambda z_s^* f_s \right)^* (\lambda x_1^*) + \left(\sum_{s \in S_2} (1 - \lambda) z_s^* f_s \right)^* ((1 - \lambda) x_2^*) < \lambda \alpha + (1 - \lambda) \beta.$$

Letting α converge towards $\eta(x_1^*)$ and β towards $\eta(x_2^*)$, one has the desired conclusion. □

The function $\operatorname{cl}_{\omega^*} \eta$ is convex and $\omega(X^*, X)$-lower semicontinuous, and its properness can be proven similarly to that of η. By applying again the Fenchel–Moreau Theorem, from (12.1) it follows that

$$\left(\sup_{t \in T} f_t \right)^* = \operatorname{cl}_{\omega^*} \left(\inf_{\substack{S \subseteq T, |S| < +\infty \\ z_s^* > 0\ \forall s \in S \\ \sum_{s \in S} z_s^* = 1}} \left(\sum_{s \in S} z_s^* f_s \right)^* \right). \tag{12.2}$$

For every arbitrary vector topology τ on X^* one has

$$\bigcup_{\substack{S\subseteq T,|S|<+\infty \\ z_s^*>0\ \forall s\in S \\ \sum_{s\in S} z_s^*=1}} \operatorname{epi}\left(\sum_{s\in S} z_s^* f_s\right)^* \subseteq \operatorname{epi}\left(\inf_{\substack{S\subseteq T,|S|<+\infty \\ z_s^*>0\ \forall s\in S \\ \sum_{s\in S} z_s^*=1}} \left(\sum_{s\in S} z_s^* f_s\right)^*\right)$$

$$\subseteq \operatorname{cl}_{\tau\times\mathcal{R}}\left(\bigcup_{\substack{S\subseteq T,|S|<+\infty \\ z_s^*>0\ \forall s\in S \\ \sum_{s\in S} z_s^*=1}} \operatorname{epi}\left(\sum_{s\in S} z_s^* f_s\right)^*\right)$$

and, consequently,

$$\operatorname{epi}\left(\sup_{t\in T} f_t\right)^* = \operatorname{cl}_{\omega^*\times\mathcal{R}}\left(\operatorname{epi}\left(\inf_{\substack{S\subseteq T,|S|<+\infty \\ z_s^*>0\ \forall s\in S \\ \sum_{s\in S} z_s^*=1}} \left(\sum_{s\in S} z_s^* f_s\right)^*\right)\right)$$

$$= \operatorname{cl}_{\omega^*\times\mathcal{R}}\left(\bigcup_{\substack{S\subseteq T,|S|<+\infty \\ z_s^*>0\ \forall s\in S \\ \sum_{s\in S} z_s^*=1}} \operatorname{epi}\left(\sum_{s\in S} z_s^* f_s\right)^*\right). \tag{12.3}$$

On the other hand, for all $x^* \in X^*$ there is

$$\left(\sup_{t\in T} f_t\right)^*(x^*) \leq \inf_{z^*\in\mathcal{P}} (z^* f)^*(x^*) \leq \inf_{\substack{S\subseteq T,|S|<+\infty \\ z_s^*>0\ \forall s\in S \\ \sum_{s\in S} z_s^*=1}} \left(\sum_{s\in S} z_s^* f_s\right)^*(x^*). \tag{12.4}$$

The first inequality is obvious, therefore, we prove only the second one. Let be $x^* \in X^*$. If the infimum in the right-hand side of the relation above is equal to $+\infty$, there is nothing to prove. Otherwise, take an arbitrary $r \in \mathbb{R}$ strictly greater than this infimum. Thus there is a finite subset S of T and some $z_s^* > 0$ for all $s \in S$ with $\sum_{s\in S} z_s^* = 1$ for which $(\sum_{s\in S} z_s^* f_s)^*(x^*) < r$. Considering a $\bar{z}^* \in \mathcal{P}$ which satisfies $\bar{z}_s^* = z_s^*$ when $s \in S$ and $\bar{z}_s^* = 0$ otherwise, it follows that $\inf_{z^*\in\mathcal{P}}(z^* f)^*(x^*) \leq (\bar{z}^* f)^*(x^*) \leq (\sum_{s\in S} z_s^* f_s)^*(x^*) < r$. Since r was arbitrarily chosen, (12.4) follows.

By taking (12.2) and (12.4) into consideration, it yields

$$\left(\sup_{t\in T} f_t\right)^*(x^*) = \mathrm{cl}_{\omega^*}\left(\inf_{z^*\in\mathcal{P}}(z^* f)^*\right)(x^*).$$

On the other hand, we have

$$\underset{z^*\in\mathcal{P}}{\cup}\operatorname{epi}(z^* f)^* \subseteq \operatorname{epi}\left(\inf_{z^*\in\mathcal{P}}(z^* f)^*\right) \subseteq \mathrm{cl}_{\omega^*\times\mathcal{R}}\left(\underset{z^*\in\mathcal{P}}{\cup}\operatorname{epi}(z^* f)^*\right).$$

The relations above lead to

$$\operatorname{epi}\left(\sup_{t\in T} f_t\right)^* = \mathrm{cl}_{\omega^*\times\mathcal{R}}\left(\operatorname{epi}\left(\inf_{z^*\in\mathcal{P}}(z^* f)^*\right)\right) = \mathrm{cl}_{\omega^*\times\mathcal{R}}\left(\underset{z^*\in\mathcal{P}}{\cup}\operatorname{epi}(z^* f)^*\right). \tag{12.5}$$

For every finite subset S of T and all $z_s^* > 0$, whenever $s \in S$ with $\sum_{s\in S} z_s^* = 1$, we have, by using Theorem 7.14(ii), $\operatorname{epi}\left(\sum_{s\in S} z_s^* f_s\right)^* = \mathrm{cl}_{\omega^*\times\mathcal{R}}\left(\sum_{s\in S}\operatorname{epi}(z_s^* f_s)^*\right)$. On the other hand, for an arbitrary proper function $k : X \to \overline{\mathbb{R}}$ and $\lambda > 0$ it holds $\operatorname{epi}(\lambda k^*) = \lambda \operatorname{epi} k^*$. Using this fact, we get further

$$\operatorname{epi}\left(\sum_{s\in S} z_s^* f_s\right)^* = \mathrm{cl}_{\omega^*\times\mathcal{R}}\left(\sum_{s\in S} z_s^* \operatorname{epi} f_s^*\right) \subseteq \mathrm{cl}_{\omega^*\times\mathcal{R}}\left(\operatorname{co}\left(\underset{t\in T}{\cup}\operatorname{epi} f_t^*\right)\right),$$

whence

$$\operatorname{epi}\left(\sup_{t\in T} f_t\right)^* \subseteq \mathrm{cl}_{\omega^*\times\mathcal{R}}\left(\operatorname{co}\left(\underset{t\in T}{\cup}\operatorname{epi} f_t^*\right)\right). \tag{12.6}$$

Taking an arbitrary element $(x^*, r) \in \operatorname{co}\left(\cup_{t\in T}\operatorname{epi} f_t^*\right)$, there is a finite subset S of T and some $z_s^* > 0$ for all $s \in S$ with $\sum_{s\in S} z_s^* = 1$ for which $(x^*, r) \in \sum_{s\in S} z_s^* \operatorname{epi} f_s^* \subseteq \operatorname{epi}\left(\sum_{s\in S} z_s^* f_s\right)^*$. This yields $\left(\sum_{s\in S} z_s^* f_s\right)^*(x^*) \le r$. Let be $\bar{z}^* \in \mathcal{P}$ which satisfies $\bar{z}_s^* = z_s^*$ when $s \in S$ and $\bar{z}_s^* = 0$, otherwise. Then there is

$$(\bar{z}^* f)^*(x^*) = \left(\sum_{s\in S} z_s^* f_s + \delta_{\underset{s\in T\setminus S}{\cap}\operatorname{dom} f_s}\right)^*(x^*) \le \left(\sum_{s\in S} z_s^* f_s\right)^*(x^*) \le r.$$

This means that $(x^*, r) \in \operatorname{epi}(\bar{z}^* f)^* \subseteq \cup_{z^*\in\mathcal{P}}\operatorname{epi}(z^* f)^*$, which yields

$$\operatorname{co}\left(\underset{t\in T}{\cup}\operatorname{epi} f_t^*\right) \subseteq \bigcup_{\substack{S\subseteq T, |S|<+\infty \\ z_s^*>0\ \forall s\in S \\ \sum_{s\in S} z_s^*=1}} \operatorname{epi}\left(\sum_{s\in S} z_s^* f_s\right)^* \subseteq \underset{z^*\in\mathcal{P}}{\cup}\operatorname{epi}(z^* f)^*. \tag{12.7}$$

By using (12.5), (12.6) and (12.7), we obtain

$$\operatorname{epi}\left(\sup_{t\in T} f_t\right)^* = \operatorname{cl}_{\omega^*\times\mathcal{R}}\left(\operatorname{co}\left(\bigcup_{t\in T}\operatorname{epi} f_t^*\right)\right). \tag{12.8}$$

On the other hand,

$$\bigcup_{t\in T}\operatorname{epi} f_t^* \subseteq \operatorname{epi}\left(\inf_{t\in T} f_t^*\right) \subseteq \operatorname{cl}_{\omega^*\times\mathcal{R}}\left(\bigcup_{t\in T}\operatorname{epi} f_t^*\right) \subseteq \operatorname{cl}_{\omega^*\times\mathcal{R}}\left(\operatorname{co}\left(\bigcup_{t\in T}\operatorname{epi} f_t^*\right)\right)$$

and from here it follows that

$$\operatorname{cl}_{\omega^*\times\mathcal{R}}\left(\operatorname{co}\left(\bigcup_{t\in T}\operatorname{epi} f_t^*\right)\right) = \operatorname{cl}_{\omega^*\times\mathcal{R}}\left(\operatorname{co}\left(\operatorname{epi}\left(\inf_{t\in T} f_t^*\right)\right)\right). \tag{12.9}$$

Here, for a function $k : X \to \overline{\mathbb{R}}$ we denote by $\operatorname{co} k : X \to \overline{\mathbb{R}}$ the *convex hull* of k, which is the greatest convex function everywhere less than or equal to k. We have that $\operatorname{epi}\operatorname{cl}(\operatorname{co} k) = \operatorname{cl}(\operatorname{co}(\operatorname{epi} k))$. Combining (12.8) with (12.9), we rediscover the known formula

$$\operatorname{epi}\left(\sup_{t\in T} f_t\right)^* = \operatorname{cl}_{\omega^*\times\mathcal{R}}\left(\operatorname{co}\left(\operatorname{epi}\left(\inf_{t\in T} f_t^*\right)\right)\right). \tag{12.10}$$

Remark 12.2. Whenever $\operatorname{co}\left(\cup_{t\in T}\operatorname{epi} f_t^*\right)$ is closed in $(X^*, \omega(X^*, X)) \times \mathbb{R}$ or, equivalently, $\operatorname{epi}(\sup_{t\in T} f_t)^* = \operatorname{co}\left(\cup_{t\in T}\operatorname{epi} f_t^*\right)$, by (12.5), (12.7) and (12.8), it yields that for all $x^* \in X^*$ it holds

$$\left(\sup_{t\in T} f_t\right)^*(x^*) = \min_{z^*\in\mathcal{P}}(z^* f)^*(x^*) = \min_{\substack{S\subseteq T, |S|<+\infty \\ z_s^*>0\ \forall s\in S \\ \sum_{s\in S} z_s^*=1}}\left(\sum_{s\in S} z_s^* f_s\right)^*(x^*).$$

Next we deal with the biconjugate function of $\sup_{t\in T} f_t$. According to [89], with a family $\mathcal{M}$ of totally saturated bounded subsets of X, one can induce on X^* the so-called *uniform convergence topology on* $\mathcal{M}$ denoted $\tau_{\mathcal{M}}$. We consider further X^* being endowed with $\tau_{\mathcal{M}}$. Restricting the sets in $\mathcal{M}$ to satisfy certain properties, one obtains different classical locally convex topologies on X^*. For instance, when all the elements of $\mathcal{M}$ are finite sets $\tau_{\mathcal{M}}$ becomes the weak* topology, when they are absolutely convex and weakly compact it coincides with the Mackey topology τ_k, while when all these sets are weakly bounded we get the strong topology on X^*, τ_b. Note that τ_b coincides, when X is a normed space, with the norm topology induced by the norm on X^*. The weak* topology is the weakest uniform convergence topology that can be considered on X^*, while τ_b is the strongest. The dual of X^* is X if and only if $\tau_{\mathcal{M}}$ is weaker than τ_k, but stronger than weak*. On the other hand, when $\tau_{\mathcal{M}}$ is strictly stronger than τ_k, but weaker than τ_b, the dual of X^*, denoted X^{**}

and referred to also as the *bidual* of X, does not coincide anymore with X. Note that by endowing X^* with any $\tau_{\mathcal{M}}$ stronger than weak* its dual has X among its linear subspaces.

In the following, we take X^* to be endowed with an arbitrary uniform convergence topology $\tau_{\mathcal{M}}$ which is strictly stronger than τ_k, but weaker than τ_b. In this case X^* has a dual X^{**} which does not coincide with X.

The following statement regarding the biconjugate of the supremum of $f_t, t \in T$, can be given (see also [21]).

Theorem 12.3. *One always has* $(\sup_{t\in T} f_t)^{**} \geq \sup_{z^*\in\mathcal{P}}(z^* f)^{**}$. *Assuming the additional hypothesis*

(H^T) f_t *is lower semicontinuous,* $t \in T$,

fulfilled, there is always

$$\left(\sup_{t\in T} f_t\right)^{**} = \left(\mathrm{cl}_{\omega^*}\left(\inf_{z^*\in\mathcal{P}}(z^* f)^*\right)\right)^* \geq \sup_{z^*\in\mathcal{P}}(z^* f)^{**}. \tag{12.11}$$

When X is a normed space the inequality in (12.11) *is always fulfilled as equality if and only if* $\mathrm{cl}_{\omega^*}(\inf_{z^*\in\mathcal{P}}(z^* f)^*) = \mathrm{cl}(\inf_{z^*\in\mathcal{P}}(z^* f)^*)$. *If* (H^T) *is fulfilled and one of the following conditions:*

(i) $\mathrm{co}\left(\bigcup_{t\in T} \mathrm{epi}\, f_t^*\right)$ *is closed in* $(X^*, \omega(X^*, X)) \times \mathbb{R}$;

(ii) T *is at most countable,* X *is a Fréchet space and*
$0 \in \mathrm{sqri}\left(\prod_{t\in T} f_t\left(\bigcap_{t\in T} \mathrm{dom}\, f_t\right) - \Delta_{\mathbb{R}^T} + \mathbb{R}^T_+\right)$;

is satisfied or if

(iii) T *is finite,*

then $(\sup_{t\in T} f_t)^{**} = \sup_{z^*\in\mathcal{P}}(z^* f)^{**}$.

Proof. For each $z^* \in \mathcal{P}$ one has $\sup_{t\in T} f_t \geq (z^* f)$ and from here $(\sup_{t\in T} f_t)^{**} \geq (z^* f)^{**}$. This yields the first inequality. Provided the fulfillment of (H_T), we have seen that one always has $(\sup_{t\in T} f_t)^* = \mathrm{cl}_{\omega^*}\left(\inf_{y^*\in\mathcal{P}}(z^* f)^*\right)$. By conjugating in both sides, we obtain

$$\left(\sup_{t\in T} f_t\right)^{**} = \left(\mathrm{cl}_{\omega^*}\left(\inf_{z^*\in\mathcal{P}}(z^* f)^*\right)\right)^* \geq \left(\inf_{z^*\in\mathcal{P}}(z^* f)^*\right)^* = \sup_{z^*\in\mathcal{P}}(z^* f)^{**}. \tag{12.12}$$

If $\mathrm{cl}_{\omega^*}(\inf_{z^*\in\mathcal{P}}(z^* f)^*) = \mathrm{cl}(\inf_{z^*\in\mathcal{P}}(z^* f)^*)$, then the inequality in (12.12) is fulfilled as equality and this means that the inequality in (12.11) turns into an equality. Viceversa, assuming that (12.11) is fulfilled as equality, it automatically follows that $(\mathrm{cl}_{\omega^*}(\inf_{z^*\in\mathcal{P}}(z^* f)^*))^* = (\mathrm{cl}(\inf_{z^*\in\mathcal{P}}(z^* f)^*))^*$ and, since X is a normed space, using a similar argumentation to the one in the proof of Theorem 10.3, we obtain $\mathrm{cl}_{\omega^*}(\inf_{z^*\in\mathcal{P}}(z^* f)^*) = \mathrm{cl}(\inf_{z^*\in\mathcal{P}}(z^* f)^*)$.

To prove the second part of the theorem, we treat each of the conditions (i)–(iii) separately. In case (H^T) is fulfilled and (i) holds, by Remark 12.2, one has that

$$\left(\sup_{t\in T} f_t\right)^* = \min_{z^*\in\mathcal{P}} (z^* f)^*. \tag{12.13}$$

We prove next that under (H^T) (ii) is also sufficient for having (12.13). We consider $C = \mathbb{R}^T_+$, $S = X \times \mathbb{R}$, $\tilde{f}(x,u) = u$ and $\tilde{g}(x,u) = f(x) - \tilde{u}$, where $\tilde{u} \in \mathbb{R}^T$ is such that $\tilde{u}_t = u$ for all $t \in T$. One can easily notice that S is nonempty, convex and closed, $\tilde{f}$ is proper, convex and lower semicontinuous, $\tilde{g}$ is proper, $\mathbb{R}^T_+$-convex and $\mathbb{R}^T_+$-epi closed and the feasibility condition $\operatorname{dom} \tilde{f} \cap S \cap \tilde{g}^{-1}(-\mathbb{R}^T_+) \neq \emptyset$ is satisfied. As T is an at most countable set, $\mathbb{R}^T$ is a Fréchet space. Thus, by Theorem 5.5, if $0 \in \operatorname{sqri}\left(\tilde{g}((\cap_{t\in T} \operatorname{dom} f_t) \times \mathbb{R}) + \mathbb{R}^T_+\right)$, which is nothing else than the regularity condition in (ii), then (as $(RC_2^{C_L})$ is fulfilled) for the convex optimization problem

$$\inf_{\substack{(x,u)\in S\\ \tilde{g}(x,u)\leqq_{\mathbb{R}^T_+} 0}} \tilde{f}(x,u) \tag{12.14}$$

and its Lagrange dual problem stable strong duality holds. Thus for all $x^* \in X^*$ it yields

$$\left(\sup_{t\in T} f_t\right)^*(x^*) = -\inf_{\substack{(x,u)\in X\times\mathbb{R}\\ f_t(x)-u\leq 0\ \forall t\in T}} \{u - \langle x^*, x\rangle\} = \min_{z^*\in(\mathbb{R}^T_+)^*} (\tilde{f} + (z^*\tilde{g}))^*(x^*, 0).$$

For all $z^* \in (\mathbb{R}^T_+)^*$ and $x^* \in X^*$ one has

$$(\tilde{f} + (z^*\tilde{g}))^*(x^*,0) = \sup_{x\in X, u\in\mathbb{R}} \left\{ \langle x^*, x\rangle - u - (z^* f)(x) + u\sum_{t\in T} z_t^* \right\} = \sup_{x\in X}\{\langle x^*, x\rangle$$

$$-(z^* f)(x)\} + \sup_{u\in\mathbb{R}} u\left(\sum_{t\in T} z_t^* - 1\right) = \begin{cases} (z^* f)^*(x^*), & \text{if } z^* \in \mathcal{P}, \\ +\infty, & \text{otherwise}. \end{cases}$$

and, therefore, the equality in (12.13) is also in this case satisfied.

Suppose now that (iii) is valid, thus let $T = \{1, \ldots, n\}$. In this case $\mathbb{R}^T$ becomes $\mathbb{R}^n$ and taking $x' \in \cap_{t=1}^n \operatorname{dom} f_t$ there is always an $u' > \max_{t=1,\ldots,n} f_t(x')$ such that $(f_t(x') - u')_{t=1,\ldots,n} \in -\operatorname{int}(\mathbb{R}^n_+)$. This means that for the convex optimization problem in (12.14) the regularity condition $(RC_1^{C_L})$ is fulfilled. Again, by Theorem 5.5, for this problem and its Lagrange dual stable strong duality holds. Thus $(\sup_{1\leq t\leq n} f_t)^* = \min\{(\sum_{t=1}^n z_t^* f_t)^* : z_t^* \geq 0, t = 1,\ldots,n, \sum_{t=1}^n z_t^* = 1\}$, which is nothing but (12.13). Alternatively, when (iii) holds one can apply [127, Corollary 2.8.11] to obtain the same conclusion.

We have proven that under any of the three conditions we considered relation (12.13) holds. Taking the conjugates of the terms in both sides of it, the desired conclusion follows. □

Remark 12.4. In a similar way, one can prove that it always holds

$$\left(\sup_{t\in T} f_t\right)^{**} \geq \sup_{\substack{S\subseteq T, |S|<+\infty \\ z_s^*>0\ \forall s\in S \\ \sum_{s\in S} z_s^*=1}} \left(\sum_{s\in S} z_s^* f_s\right)^{**}.$$

Assume that (H^T) is fulfilled. Then

$$\left(\sup_{t\in T} f_t\right)^{**} = \left(\mathrm{cl}_{\omega^*}\left(\inf_{\substack{S\subseteq T, |S|<+\infty \\ z_s^*>0\ \forall s\in S \\ \sum_{s\in S} z_s^*=1}} \left(\sum_{s\in S} z_s^* f_s\right)^{*}\right)\right)^{*} \geq \sup_{\substack{S\subseteq T, |S|<+\infty \\ z_s^*>0\ \forall s\in S \\ \sum_{s\in S} z_s^*=1}} \left(\sum_{s\in S} z_s^* f_s\right)^{**}.$$

Moreover, when X is a normed space, the inequality in the relation above is always fulfilled as equality if and only if

$$\mathrm{cl}_{\omega^*}\left(\inf_{\substack{S\subseteq T, |S|<+\infty \\ z_s^*>0\ \forall s\in S \\ \sum_{s\in S} z_s^*=1}} \left(\sum_{s\in S} z_s^* f_s\right)^{*}\right) = \mathrm{cl}\left(\inf_{\substack{S\subseteq T, |S|<+\infty \\ z_s^*>0\ \forall s\in S \\ \sum_{s\in S} z_s^*=1}} \left(\sum_{s\in S} z_s^* f_s\right)^{*}\right).$$

Assuming (H^T) and condition (i) in Theorem 12.3 fulfilled, from Remark 12.2 one has that

$$\left(\sup_{t\in T} f_t\right)^{**} = \sup_{\substack{S\subseteq T, |S|<+\infty \\ z_s^*>0\ \forall s\in S \\ \sum_{s\in S} z_s^*=1}} \left(\sum_{s\in S} z_s^* f_s\right)^{**}.$$

Remark 12.5. Taking into consideration the discussion made on the topologies that can be considered on X^*, one can expect to extend in a natural way the results given in the previous section from normed to locally convex spaces, in which case the conditions stated at (ii) would impose the spaces to be Fréchet, instead of Banach.

Concerning the biconjugate of the supremum of an (infinite) family of convex functions $(f_t)_{t\in T}$, it is natural to look for conditions which guarantee the formula $(\sup_{t\in T} f_t)^{**} = \sup_{t\in T} f_t^{**}$. In the following result, we show that under (H^T) condition (i) in Theorem 12.3 guarantees this formula, too.

Theorem 12.6. *One always has* $(\sup_{t\in T} f_t)^{**} \geq \sup_{t\in T} f_t^{**}$. *Assuming the additional hypothesis* (H_T) *and that* $\mathrm{co}\left(\bigcup_{t\in T} \mathrm{epi}\, f_t^*\right)$ *is closed in* $(X^*, \omega(X^*, X)) \times \mathbb{R}$, *it follows* $(\sup_{t\in T} f_t)^{**} = \sup_{t\in T} f_t^{**}$.

Proof. By conjugating twice both sides of the inequality $f_s \le \sup_{t\in T} f_t$ we get $f_s^{**} \le (\sup_{t\in T} f_t)^{**}$ for all $s \in T$. Thus $\sup_{t\in T} f_t^{**} \le (\sup_{t\in T} f_t)^{**}$. Since $\cup_{t\in T} \operatorname{epi} f_t^*$ is a subset of $\operatorname{epi}\left(\inf_{t\in T} f_t^*\right)$, one has

$$\operatorname{co}\left(\bigcup_{t\in T} \operatorname{epi} f_t^*\right) \subseteq \operatorname{co}\left(\operatorname{epi}\left(\inf_{t\in T} f_t^*\right)\right). \tag{12.15}$$

Since $\operatorname{epi}\left(\inf_{t\in T} f_t^*\right) \subseteq \operatorname{epi}\left(\operatorname{co}\left(\inf_{t\in T} f_t^*\right)\right)$ and the latter is a convex set, from (12.15), it yields

$$\operatorname{co}\left(\bigcup_{t\in T} \operatorname{epi} f_t^*\right) \subseteq \operatorname{co}\left(\operatorname{epi}\left(\inf_{t\in T} f_t^*\right)\right) \subseteq \operatorname{epi}\left(\operatorname{co}\left(\inf_{t\in T} f_t^*\right)\right). \tag{12.16}$$

On the other hand, by (12.8) and (12.9), there is

$$\operatorname{epi}\left(\sup_{t\in T} f_t\right)^* = \operatorname{cl}_{\omega^*\times\mathcal{R}}\left(\operatorname{co}\left(\bigcup_{t\in T} \operatorname{epi} f_t^*\right)\right) = \operatorname{cl}_{\omega^*\times\mathcal{R}}\left(\operatorname{co}\left(\operatorname{epi}\left(\inf_{t\in T} f_t^*\right)\right)\right)$$

$$= \operatorname{epi}\left(\operatorname{cl}_{\omega^*\times\mathcal{R}}\left(\operatorname{co}\left(\inf_{t\in T} f_t^*\right)\right)\right) = \operatorname{cl}_{\omega^*\times\mathcal{R}}\left(\operatorname{epi}\operatorname{co}\left(\inf_{t\in T} f_t^*\right)\right).$$

Further, as $\operatorname{co}\left(\bigcup_{t\in T} \operatorname{epi} f_t^*\right)$ is closed in $(X^*, \omega(X^*, X))\times\mathbb{R}$, we get $\operatorname{epi}(\sup_{t\in T} f_t)^* = \operatorname{epi}\left(\operatorname{co}\left(\inf_{t\in T} f_t^*\right)\right)$, which provides $(\sup_{t\in T} f_t)^* = \operatorname{co}\left(\inf_{t\in T} f_t^*\right)$. For the conjugates of these functions it holds $(\sup_{t\in T} f_t)^{**} = \left(\operatorname{co}\left(\inf_{t\in T} f_t^*\right)\right)^*$. Since

$$\operatorname{cl}\left(\operatorname{co}\left(\inf_{t\in T} f_t^*\right)\right) \le \operatorname{co}\left(\inf_{t\in T} f_t^*\right) \le \inf_{t\in T} f_t^*,$$

by [127, Theorem 2.3.1(iv)], it follows that the conjugate of $\operatorname{co}\left(\inf_{t\in T} f_t^*\right)$ coincides with the one of $\inf_{t\in T} f_t^*$. Therefore, we have $(\sup_{t\in T} f_t)^{**} = \left(\inf_{t\in T} f_t^*\right)^* = \sup_{t\in T} f_t^{**}$. □

Remark 12.7. We want to notice that we have no knowledge of a generalized interior point condition to guarantee the equality in the theorem above.

13 The Supremum of Two Convex Functions

In this section, we carry on the investigations made in Section 12 by dealing with the supremum of two convex functions. We give some refined formulae for the conjugate of the supremum and show that under the Attouch–Brézis regularity condition, one has equality between the biconjugate of the supremum and the supremum of the biconjugates of the two functions.

Consider X a normed space with X^* its topological dual space and $f, g : X \to \overline{\mathbb{R}}$ two proper and convex functions such that $\operatorname{dom} f \cap \operatorname{dom} g \neq \emptyset$. Let us denote by $f \vee g : X \to \overline{\mathbb{R}}$, $f \vee g(x) = \max\{f(x), g(x)\}$ the pointwise maximum of f and g. Being in the situation when T is finite, from the proof of Theorem 12.3 we obtain a refined version of (12.5). More precisely, for all $x^* \in X^*$ it holds

$$(f \vee g)^*(x^*) = \min_{\lambda \in [0,1]} (\lambda f + (1-\lambda)g)^* (x^*)$$

and from here one has

$$\operatorname{epi}(f \vee g)^* = \bigcup_{\lambda \in [0,1]} \operatorname{epi} (\lambda f + (1-\lambda)g)^*$$

$$= \bigcup_{\lambda \in (0,1)} \operatorname{epi} (\lambda f + (1-\lambda)g)^* \cup \operatorname{epi} \left(f + \delta_{\operatorname{dom} g}\right)^* \cup \operatorname{epi} \left(g + \delta_{\operatorname{dom} f}\right)^*. \quad (13.1)$$

Assuming further that f and g are lower semicontinuous, we get via (12.3) a second formulation for the epigraph of $\operatorname{epi}(f \vee g)^*$, namely

$$\operatorname{epi}(f \vee g)^* = \operatorname{cl}_{\omega^* \times \mathcal{R}} \left(\operatorname{epi} f^* \cup \operatorname{epi} g^* \cup \bigcup_{\lambda \in (0,1)} \operatorname{epi}(\lambda f + (1-\lambda)g)^* \right).$$

Noticing that

$$\begin{aligned} \operatorname{epi} f^* \cup \operatorname{epi} g^* &\subseteq \operatorname{cl}_{\omega^* \times \mathcal{R}} \left(\bigcup_{\lambda \in (0,1)} (\lambda \operatorname{epi} f^* + (1-\lambda) \operatorname{epi} g^*) \right) \\ &\subseteq \operatorname{cl}_{\omega^* \times \mathcal{R}} \left(\bigcup_{\lambda \in (0,1)} \operatorname{epi}(\lambda f + (1-\lambda)g)^* \right), \end{aligned}$$

it follows that $\operatorname{epi}(f \vee g)^* = \operatorname{cl}_{\omega^* \times \mathcal{R}} \left(\cup_{\lambda \in (0,1)} \operatorname{epi}(\lambda f + (1-\lambda)g)^*\right)$. On the other hand, we have

$$\begin{aligned} \bigcup_{\lambda \in (0,1)} \operatorname{epi}(\lambda f + (1-\lambda)g)^* &\subseteq \operatorname{epi} \left(\inf_{\lambda \in (0,1)} (\lambda f + (1-\lambda)g)^* \right) \\ &\subseteq \operatorname{cl}_{\omega^* \times \mathcal{R}} \left(\bigcup_{\lambda \in (0,1)} \operatorname{epi}(\lambda f + (1-\lambda)g)^* \right) \end{aligned}$$

and therefore

$$\begin{aligned} \operatorname{epi}(f \vee g)^* &= \operatorname{cl}_{\omega^* \times \mathcal{R}} \left(\inf_{\lambda \in (0,1)} (\lambda f + (1-\lambda)g)^* \right) \\ &= \operatorname{cl}_{\omega^* \times \mathcal{R}} \left(\bigcup_{\lambda \in (0,1)} \operatorname{epi}(\lambda f + (1-\lambda)g)^* \right). \end{aligned} \quad (13.2)$$

Employing the closure in $(X^*, \|\cdot\|_*) \times \mathbb{R}$, we obviously have

$$\operatorname{cl}\left(\bigcup_{\lambda\in(0,1)} \operatorname{epi}(\lambda f + (1-\lambda)g)^*\right)$$

$$\subseteq \operatorname{cl}_{\omega^*\times\mathcal{R}}\left(\bigcup_{\lambda\in(0,1)} \operatorname{epi}(\lambda f + (1-\lambda)g)^*\right) = \operatorname{epi}(f\vee g)^*.$$

Proposition 13.1. *The following inclusion always holds*

$$\operatorname{epi}\left(f+\delta_{\operatorname{dom} g}\right)^* \subseteq \operatorname{cl}\left(\bigcup_{\lambda\in(0,1)} \operatorname{epi}(\lambda f + (1-\lambda)g)^*\right).$$

Proof. Let $(x^*, r) \in \operatorname{epi}\left(f+\delta_{\operatorname{dom} g}\right)^*$ or, equivalently, $\left(f+\delta_{\operatorname{dom} g}\right)^*(x^*) \leq r$. Because g is proper, convex and lower semicontinuous it follows that g^* is a proper function and therefore there exists a $y^* \in X^*$ such that $g^*(y^*) \in \mathbb{R}$.

For all $n \geq 1$ we denote $\lambda_n := 1/n$ and $\mu_n := (n-1)/n$ and get

$$(\lambda_n g + \mu_n f)^*\left(\lambda_n y^* + \mu_n x^*\right) = \sup_{x\in X}\{\langle \lambda_n y^* + \mu_n x^*, x\rangle - \lambda_n g(x) - \mu_n f(x)\}$$

$$\leq \lambda_n \sup_{x\in\operatorname{dom} f\cap\operatorname{dom} g}\{\langle y^*, x\rangle - g(x)\} + \mu_n \sup_{x\in\operatorname{dom} f\cap\operatorname{dom} g}\{\langle x^*, x\rangle - f(x)\}$$

$$\leq \lambda_n \sup_{x\in X}\{\langle y^*, x\rangle - g(x)\} + \mu_n \sup_{x\in X}\{\langle x^*, x\rangle - (f+\delta_{\operatorname{dom} g})(x)\}$$

$$= \lambda_n g^*(y^*) + \mu_n (f+\delta_{\operatorname{dom} g})^*(x^*) \leq r + \lambda_n\left(g^*(y^*) - (f+\delta_{\operatorname{dom} g})^*(x^*)\right).$$

Thus for all $n \geq 1$ it holds $\left(\lambda_n y^* + \mu_n x^*, r + \lambda_n\left(g^*(y^*) - (f+\delta_{\operatorname{dom} g})^*(x^*)\right)\right) \in \operatorname{epi}(\lambda_n g + \mu_n f)^* \subseteq \bigcup_{\lambda\in(0,1)} \operatorname{epi}(\lambda f + (1-\lambda)g)^*$, which implies that

$$(x^*, r) \in \operatorname{cl}\left(\bigcup_{\lambda\in(0,1)} \operatorname{epi}(\lambda f + (1-\lambda)g)^*\right).$$

□

Because of the symmetry of the functions f and g, by Proposition 13.1, we also have $\operatorname{epi}\left(g+\delta_{\operatorname{dom} f}\right)^* \subseteq \operatorname{cl}\left(\bigcup_{\lambda\in(0,1)} \operatorname{epi}(\lambda f + (1-\lambda)g)^*\right)$ and so (13.1) implies that $\operatorname{epi}(f\vee g)^* \subseteq \operatorname{cl}\left(\bigcup_{\lambda\in(0,1)} \operatorname{epi}(\lambda f + (1-\lambda)g)^*\right)$. Thus (see also [42])

$$\operatorname{epi}(f\vee g)^* = \operatorname{cl}_{\omega^*\times\mathcal{R}}\left(\bigcup_{\lambda\in(0,1)} \operatorname{epi}(\lambda f + (1-\lambda)g)^*\right)$$

$$= \operatorname{cl}\left(\bigcup_{\lambda\in(0,1)} \operatorname{epi}(\lambda f + (1-\lambda)g)^*\right). \tag{13.3}$$

As the latter closure is nothing else than the closure of $\inf_{\lambda\in(0,1)}(\lambda f + (1-\lambda)g)^*$ in $(X^*, \|\cdot\|_*) \times \mathbb{R}$, it yields

$$\operatorname{epi}(f \vee g)^* = \operatorname{cl}\left(\operatorname{epi}\left(\inf_{\lambda\in(0,1)}(\lambda f + (1-\lambda)g)^*\right)\right)$$

and from here

$$(f \vee g)^* = \operatorname{cl}\left(\inf_{\lambda\in(0,1)}(\lambda f + (1-\lambda)g)^*\right). \tag{13.4}$$

In the following we turn back to the formula for $\operatorname{epi}(f \vee g)^*$ given in (13.1) and show what it becomes, provided the Attouch–Brézis regularity condition (RC_2^{id}) is fulfilled, namely X is a Banach space, f and g are lower semicontinuous and $0 \in \operatorname{sqri}(\operatorname{dom} g - \operatorname{dom} f)$. By Theorems 5.5, 7.6 and 7.7, this condition guarantees that for all $\lambda, \mu > 0$ $\operatorname{epi}(\lambda f + \mu g)^* = \operatorname{epi}(\lambda f)^* + \operatorname{epi}(\mu g)^* = \lambda \operatorname{epi} f^* + \mu \operatorname{epi} g^*$. Unfortunately, one cannot use the same argument for proving that $\operatorname{epi}(f + \delta_{\operatorname{dom} g})^* = \operatorname{epi} f^* + \operatorname{epi} \delta^*_{\operatorname{dom} g}$, as $\delta_{\operatorname{dom} g}$ is not necessarily lower semicontinuous. Nevertheless, this follows from [127, Theorem 2.8.7(v)], since both f and $\delta_{\operatorname{dom} g}$ are *li-convex* functions. Consequently, relation (13.1) becomes

$$\operatorname{epi}(f \vee g)^* = \bigcup_{\lambda\in(0,1)} (\lambda \operatorname{epi} f^* + (1-\lambda) \operatorname{epi} g^*)$$

$$\cup \left(\operatorname{epi} f^* + \operatorname{epi} \sigma_{\operatorname{dom} g}\right) \cup \left(\operatorname{epi} g^* + \operatorname{epi} \sigma_{\operatorname{dom} f}\right). \tag{13.5}$$

Therefore, if (RC_2^{id}) is fulfilled, then the conjugate of $f \vee g$ at $x^* \in X^*$ looks like

$$(f\vee g)^*(x^*) = \min\left\{ \inf_{\substack{\lambda\in(0,1), y^*, z^*\in X^* \\ \lambda y^* + (1-\lambda)z^* = x^*}} \{\lambda f^*(y^*) + (1-\lambda) g^*(z^*)\}, \min_{\substack{y^*, z^*\in X^* \\ y^*+z^*=x^*}} \{f^*(y^*)\right.$$

$$\left. + \sigma_{\operatorname{dom} g}(z^*)\}, \min_{\substack{y^*, z^*\in X^* \\ y^*+z^*=x^*}} \{g^*(y^*) + \sigma_{\operatorname{dom} f}(z^*)\} \right\}. \tag{13.6}$$

In [71, Remark 3] Fitzpatrick and Simons give an example which shows that the equality

$$(f \vee g)^*(x^*) = \min_{\substack{\lambda\in[0,1], y^*, z^*\in X^* \\ \lambda y^* + (1-\lambda)z^* = x^*}} \{\lambda f^*(y^*) + (1-\lambda) g^*(z^*)\}$$

is not true for all $x^* \in X^*$.

Regarding the biconjugate of $f \vee g$, we have seen in the previous section that in case f and g are proper, convex and lower semicontinuous fulfilling $\operatorname{dom} f \cap \operatorname{dom} g \neq \emptyset$ and $\operatorname{co}(\operatorname{epi} f^* \cup \operatorname{epi} g^*)$ is closed in $(X^*, \omega(X^*, X)) \times \mathbb{R}$, then one has on X^{**} that $(f \vee g)^{**} = f^{**} \vee g^{**}$. This is true even if X is a separated locally convex space.

We prove next, by using relation (13.4), that the same conclusion follows when X is a normed space and the Attouch–Brézis regularity condition (RC_2^{id}) holds. The proof of the next result was proposed by Constantin Zălinescu. For alternative proofs we refer to [42, 71, 129].

Theorem 13.2. *Assume that X is a Banach space and $f, g : X \to \overline{\mathbb{R}}$ are proper, convex and lower semicontinuous functions with* $\operatorname{dom} f \cap \operatorname{dom} g \neq \emptyset$ *such that* $0 \in \operatorname{sqri}(\operatorname{dom} g - \operatorname{dom} f)$. *Then $(f \vee g)^{**} = f^{**} \vee g^{**}$.*

Proof. From (13.4), by using some elementary properties of the conjugate functions, we get that

$$(f \vee g)^{**} = \left(\operatorname{cl}\left(\inf_{\lambda\in(0,1)} (\lambda f + (1-\lambda)g)^*\right)\right)^*$$

$$= \left(\inf_{\lambda\in(0,1)} (\lambda f + (1-\lambda)g)^*\right)^* = \sup_{\lambda\in(0,1)} (\lambda f + (1-\lambda)g)^{**}.$$

Assumption (ii) in Theorem 11.2 is satisfied and so for all $\lambda \in (0,1)$ it yields

$$(\lambda f + (1-\lambda)g)^{**} = (\lambda f)^{**} + ((1-\lambda)g)^{**}.$$

For all $x^{**} \in X^{**}$ and all $\lambda \in (0,1)$ we have $(\lambda f)^{**}(x^{**}) = ((\lambda f)^*)^*(x^{**}) = (\lambda f^*((1/\lambda)\cdot))^*(x^{**}) = \lambda f^{**}(x^{**})$ and, analogously, $((1-\lambda)g)^{**}(x^{**}) = (1-\lambda)g^{**}(x^{**})$. Since f, f^*, g and g^* are proper functions, f^{**} and g^{**} are also proper and we finally get

$$(f \vee g)^{**} = \sup_{\lambda\in(0,1)} \left(\lambda f^{**} + (1-\lambda)g^{**}\right) = f^{**} \vee g^{**}.$$

□

In case X is a Banach space, we formulated in this chapter two sufficient conditions that guaranteed that the biconjugate of the supremum of two proper, convex and lower semicontinuous functions was equal to the supremum of their biconjugates. More precisely, we have a generalized interior point condition and a closedness-type one. We have seen in Chapter II that, in general, the generalized interior point conditions turned to be stronger than the closedness-type one. This observation gives rise to the following open problem.

Open problem 13.3 Let X be a Banach space and $f, g : X \to \overline{\mathbb{R}}$ be proper, convex and lower semicontinuous functions with $\operatorname{dom} f \cap \operatorname{dom} g \neq \emptyset$ such that $0 \in \operatorname{sqri}(\operatorname{dom} g - \operatorname{dom} f)$. Is in this case the set $\operatorname{co}(\operatorname{epi} f^* \cup \operatorname{epi} g^*)$ necessarily closed in $(X^*, \omega(X^*, X)) \times \mathbb{R}$?

Chapter IV
Strong and Total Conjugate Duality

14 A General Closedness–Type Regularity Condition for (Only) Strong Duality

In this chapter, we are interested in formulating regularity conditions of closedness-type, which do not necessarily guarantee stable strong duality, but are sufficient for having strong duality. First, we do this for the primal–dual pair (PG)–(DG) and after that we particularize the general result to the different classes of problems investigated in the previous chapters.

Assume that X and Y are separated locally convex spaces, with X^* and Y^* their topological dual spaces, respectively, and $\Phi : X \times Y \to \overline{\mathbb{R}}$ is a proper and convex function such that $0 \in \Pr_Y(\operatorname{dom}\Phi)$. Throughout this chapter we assume that the dual spaces are endowed with the weak* topologies. As proved by Theorem 5.1, if Φ is lower semicontinuous, then for all $x^* \in X^*$ one has $(\Phi(\cdot,0))^*(x^*) = \operatorname{cl}_{\omega^*}\left(\inf_{y^* \in Y^*} \Phi^*(\cdot, y^*)\right)(x^*)$. Starting from this fact, one can formulate the following regularity condition for (PG) and its conjugate dual (DG)

(RC_5^Φ) | Φ is lower semicontinuous and $\inf_{y^* \in Y^*} \Phi^*(\cdot, y^*)$ is lower semicontinuous and exact at 0.

We say that $\inf_{y^* \in Y^*} \Phi^*(\cdot, y^*)$ is exact at $\bar{x}^* \in X^*$ if there exists $\bar{y}^* \in Y^*$ such that $\inf_{y^* \in Y^*} \Phi^*(\bar{x}^*, y^*) = \Phi^*(\bar{x}^*, \bar{y}^*)$. We have the following general result.

Theorem 14.1. *Let $\Phi : X \times Y \to \overline{\mathbb{R}}$ be a proper and convex function such that $0 \in \Pr_Y(\operatorname{dom}(\Phi))$. If (RC_5^Φ) is fulfilled, then $v(PG) = v(DG)$ and the dual (DG) has an optimal solution.*

Proof. Theorem 5.1 implies that for all $x^* \in X^*$ it holds $(\Phi(\cdot,0))^*(x^*) = \inf_{y^* \in Y^*} \Phi^*(x^*, y^*)$. Moreover, $\inf_{y^* \in Y^*} \Phi^*(\cdot, y^*)$ is exact at 0 and therefore there exists $\bar{y}^* \in Y^*$ such that (see also Theorem 1.1)

$$v(PG) = \inf_{x \in X} \Phi(x,0) = -(\Phi(\cdot,0))^*(0)$$

$$= -\Phi^*(0, \bar{y}^*) = \sup_{y^* \in Y^*} \{-\Phi^*(0, y^*)\} = v(DG)$$

and from here the conclusion follows. □

R.I. Boţ, *Conjugate Duality in Convex Optimization*, Lecture Notes in Economics and Mathematical Systems 637, DOI 10.1007/978-3-642-04900-2_4,

Remark 14.2. One should notice that under the supplementary assumption that Φ is a lower semicontinuous function, the strong duality for (PG) and (DG) can be equivalently expressed by asking that $\inf_{y^* \in Y^*} \Phi^*(\cdot, y^*)$ is *lower semicontinuous at* 0 *and exact at* 0. In this chapter, we deal with sufficient conditions for this situation and (RC_5^Φ) turns out to be an appropriate one. Moreover, one has that (RC_5^Φ) is implied by the closedness-type condition (RC_4^Φ). There are enough situations where one does not need to ensure stable strong duality, but only strong duality and in this case it is enough to ask that the condition introduced above is satisfied. Situations where (RC_5^Φ) is valid, while (RC_4^Φ) fails, will be addressed in the next sections for some particular cases of the general primal–dual pair (PG)–(DG).

First, let us furnish the formulations of (RC_5^Φ) for the case of the composed convex optimization problem (P^{CC}) and its two conjugate duals (D^{CC_1}) and (D^{CC_2}), respectively. Consider X and Z separated locally convex spaces, where Z is partially ordered by the nonempty convex cone $C \subseteq Z$, $f : X \to \overline{\mathbb{R}}$ a proper and convex function, $g : Z \to \overline{\mathbb{R}}$ a proper, convex and C-increasing function fulfilling, by convention, $g(\infty_C) = +\infty$ and $h : X \to Z^\bullet = Z \cup \{\infty_C\}$ a proper and C-convex function such that $h(\operatorname{dom} f \cap \operatorname{dom} h) \cap \operatorname{dom} g \neq \emptyset$. If, additionally, f and g are lower semicontinuous and h is star C-lower semicontinuous, then we have proved that the perturbation functions Φ^{CC_1} and Φ^{CC_2} are proper, convex and lower semicontinuous and 0 is an element of the projection of their domains on the corresponding spaces of perturbation variables. By (4.1) and (4.2), the regularity condition (RC_5^Φ) becomes in these situations

$(RC_5^{CC_1})$ | f and g are lower semicontinuous, h is star C-lower semicontinuous and $x^* \mapsto \inf_{z^* \in C^*} \{g^*(z^*) + (f + (z^*h))^*(x^*)\}$ is lower semicontinuous and exact at 0

and

$(RC_5^{CC_2})$ | f and g are lower semicontinuous, h is star C-lower semicontinuous and $x^* \mapsto \inf_{\substack{y^* \in X^* \\ z^* \in C^*}} \{g^*(z^*) + f^*(y^*) + (z^*h)^*(x^* - y^*)\}$ is lower semicontinuous and exact at 0,

respectively. From Theorem 14.1 one can derive the following result.

Theorem 14.3. *Let $f : X \to \overline{\mathbb{R}}$ be a proper and convex function, $g : Z \to \overline{\mathbb{R}}$ a proper, convex and C-increasing function fulfilling, by convention, $g(\infty_C) = +\infty$ and $h : X \to Z^\bullet = Z \cup \{\infty_C\}$ a proper and C-convex function such that $h(\operatorname{dom} f \cap \operatorname{dom} h) \cap \operatorname{dom} g \neq \emptyset$.*

(i) If $(RC_5^{CC_1})$ is fulfilled, then $v(P^{CC}) = v(D^{CC_1})$ and the dual has an optimal solution.
(ii) If $(RC_5^{CC_2})$ is fulfilled, then $v(P^{CC}) = v(D^{CC_2})$ and the dual has an optimal solution.

Remark 14.4. Under the topological assumptions made above for f, g and h, we have that (see Theorems 6.1 and 6.7)

$$(f + g \circ h)^* = \mathrm{cl}_{\omega^*}\left(\inf_{z^* \in C^*}\{g^*(z^*) + (f + (z^*h))^*(\cdot)\}\right)$$

$$= \mathrm{cl}_{\omega^*}\left(\inf_{y^* \in X^*, z^* \in C^*}\{g^*(z^*) + f^*(y^*) + (z^*h)^*(\cdot - y^*)\}\right)$$

$$\leq \inf_{z^* \in C^*}\{g^*(z^*)+(f+(z^*h))^*(\cdot)\} \leq \inf_{y^* \in X^*, z^* \in C^*}\{g^*(z^*)+f^*(y^*)+(z^*h)^*(\cdot - y^*)\}.$$

Consequently, if $x^* \mapsto \inf_{y^* \in X^*, z^* \in C^*}\{g^*(z^*) + f^*(y^*) + (z^*h)^*(x^* - y^*)\}$ is lower semicontinuous, then $x^* \mapsto \inf_{z^* \in C^*}\{g^*(z^*) + (f + (z^*h))^*(x^*)\}$ is lower semicontinuous, too. Moreover, if the first function is exact at 0, namely there exist $(\bar{y}^*, \bar{z}^*) \in X^* \times C^*$ for which

$$g^*(\bar{z}^*) + f^*(\bar{y}^*) + (\bar{z}^*h)^*(-\bar{y}^*) = \inf_{\substack{y^* \in X^* \\ z^* \in C^*}}\{g^*(z^*)+f^*(y^*) + (z^*h)^*(-y^*)\},$$

then $g^*(\bar{z}^*)+(f+(\bar{z}^*h)^*(0) = \inf_{z^* \in C^*}\{g^*(z^*)+(f+(z^*h))^*(0)\}$, which means that $x^* \mapsto \inf_{z^* \in C^*}\{g^*(z^*)+(f+(z^*h))^*(x^*)\}$ is exact at 0, too. This proves that $(RC_5^{CC_2})$ implies $(RC_5^{CC_1})$. Examples which show that the reverse implication is not true can be gathered from the ones given in the next sections for particular instances of the composed convex problem.

Remark 14.5. Similar sufficient conditions for having strong duality for (P^{CC}) and the dual problems (D^{CC_1}) and (D^{CC_2}), respectively, can be given also in case h fails to be star C-lower semicontinuous, but it is C-epi closed. For investigations made for this situation we refer to [29].

15 Strong Fenchel Duality

Consider X and Y two separated locally convex spaces, $A : X \to Y$ a linear continuous operator and $f : X \to \overline{\mathbb{R}}$ and $g : Y \to \overline{\mathbb{R}}$ proper and convex functions fulfilling $A(\operatorname{dom} f) \cap \operatorname{dom} g \neq \emptyset$. For the perturbation function Φ^A, as proved in Theorem 7.1, for all $x^* \in X^*$ we have that $\inf_{y^* \in Y^*}(\Phi^A)^*(x^*, y^*) = (f^* \Box A^*g^*)(x^*)$. By specializing (RC_5^{Φ}) we obtain a further regularity condition for (P^A) and its Fenchel dual problem (D^A)

(RC_5^A) | f and g are lower semicontinuous and $f^* \Box A^*g^*$ is lower semicontinuous and exact at 0,

while, by Theorem 14.1, we get to the following result (see also [29, 40]).

Theorem 15.1. *Let $f : X \to \overline{\mathbb{R}}$ and $g : Y \to \overline{\mathbb{R}}$ be proper and convex functions and $A : X \to Y$ be a linear continuous operator such that $A(\operatorname{dom} f) \cap \operatorname{dom} g \neq \emptyset$. If (RC_5^A) is fulfilled, then $v(P^A) = v(D^A)$ and the dual has an optimal solution.*

Coming to the situation when $X = Y$ and $A = \mathrm{id}_X$, the regularity condition (RC_5^A) turns out to be

$$(RC_5^{\mathrm{id}}) \quad \left| \begin{array}{l} f \text{ and } g \text{ are lower semicontinuous and } f^* \square g^* \text{ is lower} \\ \text{semicontinuous and exact at } 0, \end{array} \right.$$

while Theorem 15.1 furnishes the following strong duality result for $(P^{\mathrm{id}}) - (D^{\mathrm{id}})$.

Theorem 15.2. *Let $f, g : X \to \overline{\mathbb{R}}$ be proper and convex functions such that $\operatorname{dom} f \cap \operatorname{dom} g \neq \emptyset$. If (RC_5^{id}) is fulfilled, then $v(P^{\mathrm{id}}) = v(D^{\mathrm{id}})$ and the dual has an optimal solution.*

The fact that the condition above is implied by (RC_4^{id}) follows from the discussion made in Remark 14.2. In the following example, we have a situation where (RC_5^{id}) is fulfilled, unlike (RC_4^{id}).

Example 15.3 (cf. [29]). Consider the functions $f, g : \mathbb{R}^2 \to \overline{\mathbb{R}}$ defined by

$$f(x_1, x_2) = \begin{cases} 0, & \text{if } \|(x_1, x_2)\| \leq 1, \\ +\infty, & \text{otherwise} \end{cases}$$

and $g = \delta_{[1,+\infty)\times\mathbb{R}}$, where by $\|\cdot\|$ we denote the Euclidean norm of $\mathbb{R}^2$. Both functions are proper, convex and lower semicontinuous and it holds $\operatorname{dom} f \cap \operatorname{dom} g = \{(1,0)^T\}$. Thus for all $(x_1^*, x_2^*)^T \in \mathbb{R}^2$ we have $f^*(x_1^*, x_2^*) = \|(x_1^*, x_2^*)^T\|$ and

$$g^*(x_1^*, x_2^*) = \begin{cases} x_1^*, & \text{if } x_1^* \leq 0, x_2^* = 0, \\ +\infty, & \text{otherwise} \end{cases}$$

and from here

$$(f^* \square g^*)(x_1^*, x_2^*) = \inf_{(y_1^*, y_2^*)^T \in \mathbb{R}^2} \{f^*(y_1^*, y_2^*) + g^*(x_1^* - y_1^*, x_2^* - y_2^*)\}$$

$$= \inf_{\substack{(y_1^*, y_2^*)^T \in \mathbb{R}^2 \\ y_1^* \geq x_1^*, y_2^* = x_2^*}} \{\|(y_1^*, y_2^*)\| + x_1^* - y_1^*\} = x_1^* + \inf_{y_1^* \geq x_1^*} \left\{ \sqrt{(y_1^*)^2 + (x_2^*)^2} - y_1^* \right\}$$

$$= x_1^* + \inf_{y_1^* \geq x_1^*} \frac{(x_2^*)^2}{\sqrt{(y_1^*)^2 + (x_2^*)^2} + y_1^*} = x_1^*.$$

It is evident that $f^* \square g^*$ is lower-semicontinuous and, moreover, the infimum is attained at $(x_1^*, x_2^*)^T = (0,0)^T$. Thus (RC_5^{id}) is fulfilled. On the other hand, assuming that (RC_4^{id}) is true, one would have that $\operatorname{epi} f^* + \operatorname{epi} g^*$ is closed or, equivalently,

$(f + g)^* = f^* \square g^*$ and $f^* \square g^*$ is exact everywhere. Since the infimal convolution is lower semicontinuous, the first condition is fulfilled. Nevertheless, $f^* \square g^*$ is not exact at $(0, 1)^T$, as the infimum in the formula above is not attained and, consequently, (RC_4^{id}) fails.

Other examples illustrating the same fact have been given in [40] (for $f = \delta_U$ and $g = \delta_V$, where U and V are taken as in Example 6.13) and in Examples 11.1 and 11.3 in the book of Stephen Simons [120].

Returning to the initial setting, we take f identical to 0 and $g : X \to \overline{\mathbb{R}}$ a proper and convex function such that $R(A) \cap \operatorname{dom} g \neq \emptyset$. The regularity condition (RC_5^A) has the following formulation

$(RC_5^{A_g})$	g is lower semicontinuous and $A^* g^*$ is lower semicontinuous and exact at 0

and Theorem 15.1 leads to the following result.

Theorem 15.4. *Let $g : Y \to \overline{\mathbb{R}}$ be a proper and convex function and $A : X \to Y$ a linear continuous operator such that $R(A) \cap \operatorname{dom} g \neq \emptyset$. If $(RC_5^{A_g})$ is fulfilled, then $v(P^{A_g}) = v(D^{A_g})$ and the dual has an optimal solution.*

Further, considering for $m \geq 2$ $f_i : X \to \overline{\mathbb{R}}, i = 1, \ldots, m$, proper and convex functions such that $\cap_{i=1}^m \operatorname{dom} f_i \neq \emptyset$, $Y = X^m$, $g : X^m \to \overline{\mathbb{R}}$, $g(x^1, \ldots, x^m) = \sum_{i=1}^m f_i(x^i)$ and $A : X \to X^m, Ax = (x, \ldots, x)$, one obtains as regularity condition for the primal–dual pair (P^Σ)–(D^Σ)

(RC_5^Σ)	f_i is lower semicontinuous, $i = 1, \ldots, m$, $f_1^* \square \ldots \square f_m^*$ is lower semicontinuous and exact at 0,

while from Theorem 15.4 the following statement follows.

Theorem 15.5. *Let $f_i : X \to \overline{\mathbb{R}}$, $i = 1, \ldots, m$, be a proper and convex functions such that $\cap_{i=1}^m \operatorname{dom} f_i \neq \emptyset$. If (RC_5^Σ) is fulfilled, then $v(P^\Sigma) = v(D^\Sigma)$ and the dual has an optimal solution.*

In the last part of the section, we discuss the existence of the so-called *converse duality* for (P^A) and (D^A), namely the situation when the optimal objective values of the primal and dual are equal and the primal problem has an optimal solution. The approach is based on a fruitful idea used by Heinz Bauschke in [4] and later by Ng and Song in [102].

Along the hypotheses made for f and g at the beginning of the section, we assume that they are lower semicontinuous and $0 \in \operatorname{dom} f^* + A^*(\operatorname{dom} g^*)$. For the optimal objective value of the dual (D_A) we have that

$$-v(D_A) = -\sup_{y^* \in Y^*} \{-f^*(-A^* y^*) - g^*(y^*)\} = \inf_{y^* \in Y^*} \{f^*(-A^* y^*) + g^*(y^*)\}.$$

By Theorem 15.1, if an appropriate regularity condition is fulfilled, then the optimal objective value of the infimum problem in the right-hand side of the relation above is equal to the optimal objective value of its dual and the latter has an optimal solution. Since X^* and Y^* are endowed with the weak* topologies, one has that $(A^*)^* = A$, $f^{**}(x) = f(x)$ for all $x \in X$ and $g^{**}(y) = g(y)$ for all $y \in Y$. Therefore the Fenchel dual of the convex optimization problem

$$\inf_{y^* \in Y^*} \{f^*(-A^*y^*) + g^*(y^*)\} \tag{15.1}$$

is

$$\sup_{x \in X}\{-g(Ax) - f(x)\} = -\inf_{x \in X}\{f(x) + g(Ax)\} \tag{15.2}$$

and its objective value is equal to $-v(P_A)$. The feasibility assumptions $A(\text{dom} f) \cap \text{dom}\, g \neq \emptyset$ and $0 \in \text{dom} f^* + A^*(\text{dom}\, g^*)$ ensure that

$$-\infty < v(D_A) \leq v(P_A) < +\infty. \tag{15.3}$$

Further, we show how the condition (RC_5^A) looks like for the primal–dual pair of optimization problems (15.1)–(15.2). Noticing that f^* and g^* are proper, convex and lower semicontinuous functions with $-A^*(\text{dom}\, g^*) \cap \text{dom} f^* \neq \emptyset$ one has to ask for strong duality that

$$g^{**} \square (-A^{**} f^{**}) \text{ is lower semicontinuous and exact at } 0$$

or, equivalently, that

$$g \square (-Af) \text{ is lower semicontinuous and exact at } 0.$$

We have that

$$g \square (-A) f(0) = \inf_{y \in Y}\{g(y) + (-Af)(-y)\}$$

$$= \inf_{y \in Y}\left\{g(y) + \inf_{x \in X, Ax = y} f(x)\right\} = \inf_{x \in X}\{f(x) + g(Ax)\}$$

and, therefore, asking that $g \square (-Af)$ is exact at 0 is nothing else than asking that the primal problem (P^A) has an optimal solution.

Next we discuss the other assumption, namely that $g \square (-Af)$ is lower semicontinuous. For an arbitrary $y \in Y$ one has that

$$g \square (-Af)(y) = \inf_{z \in Y}\{g(z) + (-Af)(y - z)\}$$

$$= \inf_{z \in Y}\left\{g(z) + \inf_{x \in X, Ax = z - y} f(x)\right\} = \inf_{x \in X}\{f(x) + g(Ax + y)\} = \inf_{x \in X} \Phi^A(x, y).$$

Consequently, the sufficient regularity condition we derived from (RC_5^A) for having converse duality states that $h^A : Y \to \overline{\mathbb{R}}$, $h^A(y) = \inf_{x \in X} \Phi^A(x, y)$, the *infimal value function of* Φ^A is lower semicontinuous and the primal problem has an optimal solution.

This fact is not surprising at all, since converse duality follows as a direct consequence of the implication (i) $\Rightarrow$ (ii) in Theorem 1.4. Taking into consideration that $h^A(0) = v(P^A) \in \mathbb{R}$ (cf. (15.3)), one has in case h^A is lower semicontinuous that the problem (P^A) is normal and so $v(P^A) = v(D^A)$. The converse duality is in this way guaranteed.

16 Strong Lagrange and Fenchel–Lagrange Duality

In this section, we consider X and Z separated locally convex spaces, where Z is partially ordered by the nonempty convex cone $C \subseteq Z$, $S \subseteq X$ a nonempty and convex set, $f : X \to \overline{\mathbb{R}}$ a proper and convex function and $g : X \to Z^\bullet$ a proper and C-convex function such that $\operatorname{dom} f \cap S \cap g^{-1}(-C) \neq \emptyset$. For the primal optimization problem with geometric and cone constraints (P^C) we introduced in Section 3 via perturbation three conjugate dual problems. The general results given in Section 14 allow the formulation of weak regularity conditions for these primal–dual pairs. Considering $\Phi^{C_F} : X \times X \to \overline{\mathbb{R}}$, $\Phi^{C_F}(x, y) = f(x + y) + \delta_{\mathcal{A}}(x)$ as perturbation function, one obtains as a direct particularization of (RC_5^{id}) the following regularity condition

$$(RC_5^{C_F}) \left| \begin{array}{l} f \text{ is lower semicontinuous, } \mathcal{A} \text{ is closed and } f^*\Box\sigma_{\mathcal{A}} \text{ is} \\ \text{lower semicontinuous and exact at } 0, \end{array} \right.$$

while Theorem 15.2 furnishes the following strong duality statement.

Theorem 16.1. *Let* $S \subseteq X$ *be a nonempty convex set,* $f : X \to \overline{\mathbb{R}}$ *a proper and convex function and* $g : X \to Z^\bullet$ *a proper and* C*-convex function such that* $\operatorname{dom} f \cap S \cap g^{-1}(-C) \neq \emptyset$. *If* $(RC_5^{C_F})$ *is fulfilled, then* $v(P^C) = v(D^{C_F})$ *and the dual has an optimal solution.*

The Lagrange dual to (P^C) has been introduced in Section 3 via the perturbation function $\Phi^{C_L} : X \times Z \to \overline{\mathbb{R}}$, $\Phi^{C_L}(x, z) = f(x) + \delta_{\{u \in S : g(u) \in z - C\}}(x)$. If, additionally, S is closed, f is lower semicontinuous and g is C-epi closed, then Φ^{C_L} is lower semicontinuous, too, and from (RC_5^{Φ}) and Theorem 14.1 via (3.1), we obtain a further regularity condition

$$(RC_5^{C_L}) \left| \begin{array}{l} S \text{ is closed, } f \text{ is lower semicontinuous, } g \text{ is } C\text{-epi closed and} \\ x^* \mapsto \inf\limits_{z^* \in C^*} (f + (z^*g) + \delta_S)^*(x^*) \text{ is lower semicontinuous and} \\ \text{exact at } 0, \end{array} \right.$$

respectively, a further strong duality theorem for Lagrange duality (see also [24]).

Theorem 16.2. *Let* $S \subseteq X$ *be a nonempty convex set,* $f : X \to \overline{\mathbb{R}}$ *a proper and convex function and* $g : X \to Z^\bullet$ *a proper and* C*-convex function such that*

$\mathrm{dom}\, f \cap S \cap g^{-1}(-C) \neq \emptyset$. *If* $(RC_5^{C_L})$ *is fulfilled, then* $v(P^C) = v(D^{C_L})$ *and the dual has an optimal solution.*

Remark 16.3. In [86] Jeyakumar, Song, Dinh and Lee considered in the hypotheses that X is a Fréchet space, S is closed, $f : X \to \mathbb{R}$ is a continuous function and g is a star C-lower semicontinuous function as regularity condition for (P^C) and its Lagrange dual a sufficient condition called *dual CQ* (dCQ). The set $\mathcal{A}$ is said to fulfill (dCQ) if

$$\bigcup_{z^* \in C^*} \mathrm{epi}((z^*g) + \delta_S)^* \text{ is closed in } (X^*, \omega(X^*, X)) \times \mathbb{R}.$$

From Theorem 8.2 one has in this situation that $\mathrm{epi}\, \sigma_{\mathcal{A}} = \cup_{z^* \in C^*} \mathrm{epi}((z^*g) + \delta_S)^*$ and from Theorems 5.5, 7.6 and 7.7, it yields

$$\mathrm{epi}(f + \delta_{\mathcal{A}})^* = \mathrm{cl}_{\omega^* \times \mathcal{R}} \left(\bigcup_{z^* \in C^*} \mathrm{epi}(f + (z^*g) + \delta_S)^* \right) = \mathrm{epi}\, f^* + \mathrm{epi}\, \sigma_{\mathcal{A}}$$

$$= \mathrm{epi}\, f^* + \bigcup_{z^* \in C^*} \mathrm{epi}((z^*g) + \delta_S)^* \subseteq \bigcup_{z^* \in C^*} \mathrm{epi}(f + (z^*g) + \delta_S)^*,$$

which further implies that $\cup_{z^* \in C^*} \mathrm{epi}(f + (z^*g) + \delta_S)^*$ is closed in $(X^*, \omega(X^*, X)) \times \mathbb{R}$. Therefore, in the hypotheses considered in [86] (whereby X can be taken as a general separated locally convex space) one has not only strong duality for (P^C) and (D^{C_L}), but also stable strong duality. The same happens when $f : X \to \mathbb{R}$ is continuous, $g : X \to Z$ is continuous and $(CCCQ)$ holds, namely $\cup_{z^* \in C^*} \mathrm{epi}(z^*g)^* + \mathrm{epi}\, \sigma_S$, which is in this case equal to $\cup_{z^* \in C^*} \mathrm{epi}((z^*g) + \delta_S)^*$, is closed in $(X^*, \omega(X^*, X)) \times \mathbb{R}$. This second condition has been considered by Jeyakumar, Dinh and Lee in [84]. Consequently, both conditions are sufficient for $(RC_5^{C_L})$ and there might occur situations, like the one in the following example, where they fail, while the latter is fulfilled.

Example 16.4. Considering $X = Z = \mathbb{R}$, $S = \mathbb{R}$, $C = \mathbb{R}_+$, $f(x) = 0$ and $g(x) = x^2$, $x \in \mathbb{R}$, we have that (see also Example 8.12)

$$\bigcup_{z^* \in C^*} \mathrm{epi}(f + (z^*g) + \delta_S)^* = \mathrm{epi}\, f^* + \bigcup_{z^* \in C^*} \mathrm{epi}((z^*g) + \delta_S)^* = \{0\} \times \mathbb{R}_+ \cup \mathbb{R} \times (0, +\infty),$$

which is obviously not a closed set. Thus $(RC_4^{C_L})$ as well as the sufficient conditions in [84, 86] fail.

For $z^* = 0$ it holds $(f + (z^*g) + \delta_S)^* = \delta_{\{0\}}$ while for $z^* > 0$ and $x^* \in \mathbb{R}$ it yields $(f + (z^*g) + \delta_S)^*(x^*) = \sup_{x \in \mathbb{R}}\{x^*x - z^*x^2\} = (x^*)^2/(4z^*)$. This means that for all $x^* \in \mathbb{R}$, $\inf_{z^* \geq 0}(f + (z^*g) + \delta_S)^*(x^*) = 0$ and this infimum is attained at $x^* = 0$. The regularity condition $(RC_5^{C_L})$ holds and so we have a situation which proves that this condition is weaker than $(RC_4^{C_L})$.

Taking as perturbation function $\Phi^{C_{FL}} : X \times X \times Z \to \overline{\mathbb{R}}$, $\Phi^{C_{FL}}(x, y, z) = f(x + y) + \delta_{\{u \in S: g(u) \in z - C\}}(x)$, one can derive from (RC_5^{Φ}) and by using (3.3) the following sufficient condition for strong duality for (P^C) and its Fenchel–Lagrange dual

$$(RC_5^{C_{FL}}) \left| \begin{array}{l} S \text{ is closed, } f \text{ is lower semicontinuous, } g \text{ is } C\text{-epi closed and} \\ x^* \mapsto \inf\limits_{\substack{y^* \in X^* \\ z^* \in C^*}} \{f^*(y^*) + ((z^*g) + \delta_S)^*(x^* - y^*)\} \text{ is lower} \\ \text{semicontinuous and exact at } 0, \end{array} \right.$$

while Theorem 14.1 furnishes the following strong duality result (see also [24]).

Theorem 16.5. *Let $S \subseteq X$ be a nonempty convex set, $f : X \to \overline{\mathbb{R}}$ a proper and convex function and $g : X \to Z^\bullet$ a proper and C-convex function such that* $\operatorname{dom} f \cap S \cap g^{-1}(-C) \neq \emptyset$. *If $(RC_5^{C_{FL}})$ is fulfilled, then $v(P^C) = v(D^{C_{FL}})$ and the dual has an optimal solution.*

Remark 16.6. The fact that $(RC_4^{C_{FL}})$ implies $(RC_5^{C_{FL}})$ follows from the considerations made in Remark 14.2. It is enough to consider the sets and the functions as in the example above in order to see that the opposite implication is not true.

Remark 16.7. By taking into consideration (3.4) and Remark 3.7, it follows that whenever $(RC_5^{C_{FL}})$ is fulfilled, then one has $v(P^C) = v(D^{C_L}) = v(D^{C_F}) = v(D^{C_{FL}})$ and the dual problems have optimal solutions. Another sufficient condition for having strong duality for (P^C) and $(D^{C_{FL}})$ (as we have seen, this implies the existence of strong duality for the other two primal–dual pairs) is the one given in [39, Theorem 4.5], namely by asking that X is a Banach space, S a closed set, f a lower semicontinuous function, $g : X \to Z$ a continuous functions, $f^* \square \sigma_{\mathcal{A}}$ is lower semicontinuous and exact at 0 and $\cup_{z^* \in C^*} \operatorname{epi}(z^*g)^* + \operatorname{epi} \sigma_S$ is closed in $(X^*, \omega(X^*, X)) \times \mathbb{R}$.

We replace the hypotheses of [39, Theorem 4.5] by some weaker conditions, by taking X a separated locally convex space, S a closed set, f a lower semicontinuous function and g a C-epi closed function such that

(i) $f^* \square \sigma_{\mathcal{A}}$ is lower semicontinuous and exact at 0
(ii) $\cup_{z^* \in C^*} \operatorname{epi}((z^*g) + \delta_S)^*$ is closed in $(X^*, \omega(X^*, X)) \times \mathbb{R}$

and prove that even under these weak hypotheses $(RC_5^{C_{FL}})$ holds. One can easily notice that if g is continuous, then $\cup_{z^* \in C^*} \operatorname{epi}(z^*g)^* + \operatorname{epi} \sigma_S = \cup_{z^* \in C^*} \operatorname{epi}((z^*g) + \delta_S)^*$ and this proves in fact that we are here in a weaker setting than in [39, Theorem 4.5].

Indeed, by Theorems 7.6, 8.1 and 8.3 we have that for all $x^* \in X^*$

$$(f + \delta_{\mathcal{A}})^*(x^*) = \operatorname{cl}_{\omega^*}\left(\inf_{\substack{y^* \in X^* \\ z^* \in C^*}} \{f^*(y^*) + ((z^*g) + \delta_S)^*(x^* - y^*)\} \right)$$

$$= \operatorname{cl}_{\omega^*}(f^* \square \sigma_{\mathcal{A}})(x^*) = (f^* \square \sigma_{\mathcal{A}})(x^*) = \inf_{y^* \in X^*} \{f^*(y^*) + \sigma_{\mathcal{A}}(x^* - y^*)\}$$

$$= \inf_{y^* \in X^*} \left\{ f^*(y^*) + \inf_{z^* \in C^*} ((z^*g) + \delta_S)^*(x^* - y^*) \right\}$$

$$= \inf_{\substack{y^* \in X^* \\ z^* \in C^*}} \{ f^*(y^*) + ((z^*g) + \delta_S)^*(x^* - y^*) \},$$

which means that $x^* \mapsto \inf_{y^* \in X^*, z^* \in C^*} \{ f^*(y^*) + ((z^*g) + \delta_S)^*(x^* - y^*) \}$ is lower semicontinuous. Moreover, since $f^* \square \sigma_{\mathcal{A}}$ is exact at 0, there exists $\bar{y}^* \in X^*$ such that $(f + \delta_{\mathcal{A}})^*(0) = f^*(\bar{y}^*) + \sigma_{\mathcal{A}}(-\bar{y}^*)$. By Theorem 8.3, using that (ii) is fulfilled, there exists $\bar{z}^* \in C^*$ such that

$$\sigma_{\mathcal{A}}(-\bar{y}^*) = \delta^*_{\mathcal{A}}(-\bar{y}^*) = -\inf_{x \in S} \{ \langle \bar{y}^*, x \rangle + (\bar{z}^*g)(x) \} = ((\bar{z}^*g) + \delta_S)^*(-\bar{y}^*)$$

and so

$$(f + \delta_{\mathcal{A}})^*(0) = \inf_{\substack{y^* \in X^* \\ z^* \in C^*}} \{ f^*(y^*) + ((z^*g) + \delta_S)^*(-y^*) \} = f^*(\bar{y}^*) + ((\bar{z}^*g) + \delta_S)^*(-\bar{y}^*).$$

Consequently, $x^* \mapsto \inf_{y^* \in X^*, z^* \in C^*} \{ f^*(y^*) + ((z^*g) + \delta_S)^*(x^* - y^*) \}$ is exact at 0 and $(RC_5^{C_{FL}})$ is satisfied.

For the problem treated in Example 16.4 we already have seen that $(RC_5^{C_{FL}})$ holds. Obviously, $f^* \square \sigma_{\mathcal{A}}$ is lower semicontinuous and exact at 0. Nevertheless, since $\cup_{z^* \in C^*} \operatorname{epi}((z^*g) + \delta_S)^* = \{0\} \times \mathbb{R}_+ \cup \mathbb{R} \times (0, +\infty)$, this set fails to be closed, thus the assumption (ii) above is not fulfilled and one can conclude that $(RC_5^{C_{FL}})$ is stronger than the condition used in [39, Theorem 4.5].

Remark 16.8. Concerning the relations between the closedness-type regularity conditions introduced in this section for the problem with cone and geometric constraints, one can show that $(RC_5^{C_{FL}})$ is weaker than $(RC_5^{C_F})$ and $(RC_5^{C_L})$.

Indeed, assume $(RC_5^{C_{FL}})$ fulfilled. We have

$$(f + \delta_{\mathcal{A}})^* = \operatorname{cl}_{\omega^*}(f^* \square \sigma_{\mathcal{A}}) = \inf_{\substack{y^* \in X^* \\ z^* \in C^*}} \{ f^*(y^*) + ((z^*g) + \delta_S)^*(\cdot - y^*) \}$$

$$\geq \inf_{y^* \in X^*} \{ f^*(y^*) + \sigma_{\mathcal{A}}(\cdot - y^*) \} = f^* \square \sigma_{\mathcal{A}}$$

and, therefore, $f^* \square \sigma_{\mathcal{A}}$ is lower semicontinuous. At $x^* = 0$ we have that there exists $(\bar{y}^*, \bar{z}^*) \in X^* \times C^*$ with

$$(f + \delta_{\mathcal{A}})^*(0) = f^*(\bar{y}^*) + ((\bar{z}^*g) + \delta_S)^*(-\bar{y}^*) \geq f^*(\bar{y}^*) + \sigma_{\mathcal{A}}(-\bar{y}^*) \geq (f + \delta_{\mathcal{A}})^*(0)$$

and this provides the exactness of $f^* \square \sigma_{\mathcal{A}}$ at 0. Therefore, $(RC_5^{C_{FL}}) \Rightarrow (RC_5^{C_F})$. For the problem in Example 8.12, we have proved that $(RC_4^{C_F})$ is satisfied and, consequently, $(RC_5^{C_F})$ is satisfied, too. Let be $x^*, y^* \in \mathbb{R}$. For $z^* = 0$ one has that $f^*(y^*) + ((z^*g) + \delta_S)^*(x^* - y^*)$ is equal to 0 if $x^* = y^* = 1$, being

$+\infty$, otherwise. For $z^* > 0$ it holds $f^*(y^*) + ((z^*g) + \delta_S)^*(x^* - y^*) = (x^* - 1)^2/4z^*$ if $y^* = 1$, being also equal to $+\infty$, otherwise. Thus the function $x^* \mapsto \inf_{y^* \in X^*, z^* \geq 0}\{f^*(y^*) + ((z^*g) + \delta_S)^*(x^* - y^*)\}$, which is identical to 0, is lower semicontinuous, but it is not exact at 0. In conclusion, the implication proved above does not hold as equivalence.

We come now to the investigation of the relationship between $(RC_5^{C_{FL}})$ and $(RC_5^{C_L})$. Since (cf. Theorem 8.1)

$$(f + \delta_{\mathcal{A}})^* = \text{cl}_{\omega^*}\left(\inf_{z^* \in C^*}(f + (z^*g) + \delta_S)^*(\cdot)\right)$$

$$= \inf_{\substack{y^* \in X^* \\ z^* \in C^*}}\{f^*(y^*) + ((z^*g) + \delta_S)^*(\cdot - y^*)\} \geq \inf_{z^* \in C^*}(f + (z^*g) + \delta_S)^*(\cdot),$$

the mapping $x^* \mapsto \inf_{z^* \in C^*}(f + (z^*g) + \delta_S)^*(x^*)$ turns out to be lower semicontinuous. At $x^* = 0$, we have that there exists $(\bar{y}^*, \bar{z}^*) \in X^* \times C^*$ with

$$(f+\delta_{\mathcal{A}})^*(0) = f^*(\bar{y}^*)+((\bar{z}^*g)+\delta_S)^*(-\bar{y}^*) \geq (f+(\bar{z}^*g)+\delta_S)^*(0) \geq (f+\delta_{\mathcal{A}})^*(0)$$

and this provides the exactness of $x^* \mapsto \inf_{z^* \in C^*}(f + (z^*g) + \delta_S)^*(x^*)$ at 0. Therefore, $(RC_5^{C_{FL}}) \Rightarrow (RC_5^{C_L})$. Denote by $\|\cdot\|$ the Euclidean norm on $\mathbb{R}^2$ and take $X = \mathbb{R}^2$, $Z = \mathbb{R}$, $C = \mathbb{R}_+$, $S = \{(x_1, x_2)^T \in \mathbb{R}^2 : \|(x_1, x_2)^T\| \leq 1\}$, $f(x_1, x_2) = \delta_{[1,+\infty)\times\mathbb{R}}(x_1, x_2) + x_2$, $(x_1, x_2)^T \in \mathbb{R}^2$, and g identical to 0. Then $\inf_{z^* \geq 0}(f + (z^*g) + \delta_S)^* = (f + \delta_S)^*$ and this function is lower semicontinuous and exact everywhere, which means that $(RC_4^{C_L})$ and, consequently, $(RC_5^{C_L})$ are fulfilled. On the other hand, for all $(x_1^*, x_2^*)^T \in \mathbb{R}^2$ it yields (see also Example 15.3)

$$\inf_{(y_1^*, y_2^*)^T \in \mathbb{R}^2, z^* \geq 0}\{f^*(y_1^*, y_2^*) + ((z^*g) + \delta_S)^*(x_1^* - y_1^*, x_2^* - y_2^*)\}$$

$$= \inf_{\substack{(y_1^*, y_2^*)^T \in \mathbb{R}^2 \\ y_1^* \leq 0, y_2^* = 1}}\left\{y_1^* + \sqrt{(x_1^* - y_1^*)^2 + (x_2^* - y_2^*)^2}\right\}$$

$$= x_1^* + \inf_{v_1^* \geq x_1^*}\frac{(x_2^* - 1)^2}{\sqrt{(v_1^*)^2 + (x_2^* - 1)^2} + v_1^*} = x_1^*.$$

Even if $(x_1^*, x_2^*) \mapsto \inf_{(x_1^*, x_2^*)^T \in \mathbb{R}^2, z^* \geq 0}\{f^*(y_1^*, y_2^*) + ((z^*g) + \delta_S)^*(x_1^* - y_1^*, x_2^* - y_2^*)\}$ is lower semicontinuous, it is not exact at $(0, 0)^T$ and therefore $(RC_5^{C_{FL}})$ fails in this case, too.

In the remaining of the section, we equivalently characterize the fulfillment of the condition on the feasible set $\mathcal{A}$ introduced in Remark 16.3 under the name *dual* (CQ) (dCQ). To this end, we additionally assume that S is a closed set and g is a C-epi closed function such that $\mathcal{A}$ is nonempty. One should notice that it is enough to suppose that g is continuous at some point of $\mathcal{A}$ (even if one renounces at the topological assumptions from above) in order to have that (dCQ) coincides with $(CCCQ)$. The following assertion is a direct consequence of Theorem 8.3.

Theorem 16.9. *The set $\mathcal{A}$ fulfills (dCQ) if and only if for all $x^* \in X^*$ one has* $\inf_{x\in\mathcal{A}}\langle x^*, x\rangle = \max_{z^*\in C^*}\inf_{x\in S}\{\langle x^*, x\rangle + (z^*g)(x)\}$.

We use this result for proving the following statement (see [31]).

Theorem 16.10. *The set $\mathcal{A}$ fulfills (dCQ) if and only if for each proper, convex and lower semicontinuous function $f : X \to \overline{\mathbb{R}}$ which satisfies* $\operatorname{dom} f \cap \mathcal{A} \neq \emptyset$ *and $(RC_5^{C_F})$ one has that $v(P^C) = v(D^{C_L}) = v(D^{C_F}) = v(D^{C_{FL}})$ and the dual problems have optimal solutions.*

Proof. The sufficiency follows from the previous theorem by taking f linear and continuous. Notice that every linear continuous function satisfies $(RC_5^{C_F})$.

For proving the necessity we consider a function f, which fulfills the hypotheses. Then there exists $\bar{y}^* \in X^*$ such that $(f + \delta_{\mathcal{A}})^*(0) = (f^*\square\sigma_{\mathcal{A}})(0) = f^*(\bar{y}^*) + \sigma_{\mathcal{A}}(-\bar{y}^*)$. Now, by Theorem 16.9, there is a $\bar{z}^* \in C^*$ fulfilling

$$\sigma_{\mathcal{A}}(-\bar{y}^*) = -\inf_{x\in S}\{\langle \bar{y}^*, x\rangle + (\bar{z}^*g)(x)\} = ((\bar{z}^*g) + \delta_S)^*(-\bar{y}^*)$$

and so $v(P^C) = -(f + \delta_{\mathcal{A}})^*(0) = -f^*(\bar{y}^*) - ((\bar{z}^*g) + \delta_S)^*(-\bar{y}^*) = v(D^{C_{FL}})$. The conclusion follows via Remark 3.7. □

Remark 16.11. In Theorem 16.10 instead of $(RC_5^{C_F})$ one can consider each of the regularity conditions $(RC_i^{C_F}), i \in \{1, 2, 3, 4\}$.

Remark 16.12. The *Farkas–Minkowski property* of a system of (infinitely many) convex or linear inequalities has been treated in the literature dealing with semi-infinite programming problems (see, for instance [64,74,75,82]) and turns out to be a special case of (dCQ).

Indeed, let T be an arbitrary index set and consider a family of functions $g_t : X \to \mathbb{R}, t \in T$, which are convex and continuous at some point of $\{x \in S : g_t(x) \leq 0, t \in T\}$. Take $C = \mathbb{R}_+^T$ and define $g : X \to \mathbb{R}^T$ as being $g(x) = (g_t(x))_{t\in T}$. We note that in this setting g does not need to be $\mathbb{R}_+^T$-epi closed. Given these, the condition (dCQ) becomes equivalent to saying that

$$\operatorname{cone}\operatorname{co}(\cup_{t\in T}\operatorname{epi} g_t^*) + \operatorname{epi}\sigma_S \text{ is closed in } (X^*, \omega(X^*, X)) \times \mathbb{R},$$

which is actually the *Farkas–Minkowski condition* (FM) in [74]. Indeed, for each $z^* \in (\mathbb{R}_+^T)^*$ one has, by Theorems 5.5, 7.6 and 7.7 $\operatorname{epi}((z^*g) + \delta_S)^* = \sum_{t\in T} z_t^* \operatorname{epi} g_t^* + \operatorname{epi}\sigma_S$. Further,

$$\bigcup_{z^*\in(\mathbb{R}_+^T)^*}\operatorname{epi}((z^*g)+\delta_S)^* = \left\{\sum_{t\in T'} z_t^* \operatorname{epi} g_t^* : T' \subseteq T, |T'| < +\infty, z_t^* > 0, t \in T'\right\}$$

$$\cup\{0\} \times \mathbb{R}_+ + \operatorname{epi} \sigma_S = \operatorname{coneco}\left(\bigcup_{t \in T} \operatorname{epi} g_t^* \cup \{(0,1)\} + \operatorname{epi} \sigma_S \right)$$

$$= \operatorname{coneco}\left(\bigcup_{t \in T} \operatorname{epi} g_t^* \right) + \operatorname{epi} \sigma_S, \text{ since } \{0\} \times \mathbb{R}_+ \subseteq \operatorname{epi} \sigma_S.$$

The following theorem provides optimality conditions for the problem (P^C). For more results of this kind we refer the reader to [31].

Theorem 16.13. *If $\mathcal{A}$ fulfills the condition (dCQ) and $f : X \to \overline{\mathbb{R}}$ is a proper, convex and lower semicontinuous function which satisfies $(RC_5^{C_F})$, then $\bar{x} \in \operatorname{dom} f \cap \mathcal{A}$ is an optimal solution to (P^C) if and only if there is some $\bar{z}^* \in C^*$ such that $(\bar{z}^* g)(\bar{x}) = 0$ and $0 \in \partial f(\bar{x}) + \partial((\bar{z}^* g) + \delta_S)(\bar{x})$.*

Proof. Since (dCQ) holds, then by Theorem 8.6 (i) $\Rightarrow$ (ii), one has

$$\partial \delta_{\mathcal{A}}(x) = \bigcup_{\substack{z^* \in C^* \\ (z^* g)(x) = 0}} \partial((z^* g) + \delta_S)(x) \ \forall x \in \mathcal{A}.$$

We know that $\bar{x} \in \operatorname{dom} f \cap \mathcal{A}$ is an optimal solution to (P^C) if and only if $0 \in \partial(f + \delta_{\mathcal{A}})(\bar{x})$. The necessity follows by taking into consideration that we always have

$$\partial f(\bar{x}) + \bigcup_{\substack{z^* \in C^* \\ (z^* g)(\bar{x}) = 0}} \partial((z^* g) + \delta_S)(\bar{x}) \subseteq \partial(f + \delta_{\mathcal{A}})(\bar{x}).$$

For the sufficiency we use $(RC_5^{C_F})$ and obtain an element $\bar{y}^* \in X^*$ such that $(f + \delta_{\mathcal{A}})^*(0) = f^* \square \sigma_{\mathcal{A}}(0) = f^*(\bar{y}^*) + \sigma_{\mathcal{A}}(-\bar{y}^*)$. Further, we get

$$0 = f(\bar{x}) + (f + \delta_{\mathcal{A}})^*(0) = f(\bar{x}) + \delta_{\mathcal{A}}(\bar{x}) + f^*(\bar{y}^*) + \sigma_{\mathcal{A}}(-\bar{y}^*) \geq 0$$

and thus one must have equalities in the Young–Fenchel inequalities for the pairs f and f^* and $\delta_{\mathcal{A}}$ and $\sigma_{\mathcal{A}}$, respectively, i.e. $\bar{y}^* \in \partial f(\bar{x})$ and $-\bar{y}^* \in \partial \delta_{\mathcal{A}}(\bar{x})$. Thus there exists $\bar{z}^* \in C^*$ fulfilling $(\bar{z}^* g)(\bar{x}) = 0$ and $-\bar{y}^* \in \partial((\bar{z}^* g) + \delta_S)(\bar{x})$ and the conclusion follows. □

17 Total Lagrange and Fenchel–Lagrange Duality

Next, we deal with another instance of strong duality for the problem (P^C) and its Lagrange and Fenchel–Lagrange duals, namely the situation when additionally an optimal solution of the primal problem is assumed to be known. We call this situation *total duality*. We keep on working in the setting from the previous section and prove for the beginning the following result (see also [31, Theorem 3]).

Theorem 17.1. *Let be* $\bar{x} \in \operatorname{dom} f \cap \mathcal{A}$. *Then*

$$\partial(f + \delta_{\mathcal{A}})(\bar{x}) = \bigcup_{\substack{z^* \in C^* \\ (z^* g)(\bar{x}) = 0}} \partial(f + (z^* g) + \delta_S)(\bar{x}) \tag{17.1}$$

if and only if for each $x^* \in X^*$ *for which the infimum over* $\mathcal{A}$ *of the function* $f + \langle x^*, \cdot \rangle$ *is attained at* $\bar{x}$ *one has*

$$f(\bar{x}) + \langle x^*, \bar{x} \rangle = \min_{x \in \mathcal{A}} \{f(x) + \langle x^*, x \rangle\} = \max_{z^* \in C^*} \inf_{x \in S} \{f(x) + \langle x^*, x \rangle + (z^* g)(x)\}. \tag{17.2}$$

Proof. Let $\bar{x} \in \operatorname{dom} f \cap \mathcal{A}$ be fixed.

"$\Rightarrow$" Take $x^* \in X^*$ such that $\bar{x}$ solves $\inf_{x \in \mathcal{A}} \{f(x) + \langle x^*, x \rangle\}$. Thus $0 \in \partial(f + \langle x^*, \cdot \rangle + \delta_{\mathcal{A}})(\bar{x})$ or, equivalently, $-x^* \in \partial(f + \delta_{\mathcal{A}})(\bar{x})$. Because (17.1) is satisfied at $\bar{x}$, there is some $\bar{z}^* \in C^*$ such that $(\bar{z}^* g)(\bar{x}) = 0$ and $-x^* \in \partial(f + (\bar{z}^* g) + \delta_S)(\bar{x}) \Leftrightarrow 0 \in \partial(f + \langle x^*, \cdot \rangle + (\bar{z}^* g) + \delta_S)(\bar{x})$, which is the same with

$$f(\bar{x}) + \langle x^*, \bar{x} \rangle = f(\bar{x}) + \langle x^*, \bar{x} \rangle + (\bar{z}^* g)(\bar{x}) = \inf_{x \in S} \{f(x) + \langle x^*, x \rangle + (\bar{z}^* g)(x)\}.$$

Because the inequality

$$\inf_{x \in \mathcal{A}} \{f(x) + \langle x^*, x \rangle\} \geq \sup_{z^* \in C^*} \inf_{x \in S} \{f(x) + \langle x^*, x \rangle + (z^* g)(x)\}$$

is always fulfilled, we get (17.2).

"$\Leftarrow$" Since the inclusion

$$\bigcup_{\substack{z^* \in C^* \\ (z^* g)(\bar{x}) = 0}} \partial(f + (z^* g) + \delta_S)(\bar{x}) \subseteq \partial(f + \delta_{\mathcal{A}})(\bar{x})$$

holds always, we have to prove only the reversed one. Take $x^* \in \partial(f + \delta_{\mathcal{A}})(\bar{x})$. This means that $\bar{x}$ solves the problem $\inf_{x \in \mathcal{A}} \{f(x) - \langle x^*, x \rangle\}$. By (17.2) there is some $\bar{z}^* \in C^*$ such that $f(\bar{x}) - \langle x^*, \bar{x} \rangle = \inf_{x \in S} \{f(x) - \langle x^*, x \rangle + (\bar{z}^* g)(x)\}$. As the infimum in the right-hand side is less than or equal to $f(\bar{x}) - \langle x^*, \bar{x} \rangle + (\bar{z}^* g)(\bar{x})$, we get that $(\bar{z}^* g)(\bar{x}) \geq 0$. Thus, because $\bar{x} \in \mathcal{A}$ and $\bar{z}^* \in C^*$, we have $(\bar{z}^* g)(\bar{x}) = 0$. This yields

$$f(\bar{x}) - \langle x^*, \bar{x} \rangle + (\bar{z}^* g)(\bar{x}) = \inf_{x \in S} \{f(x) - \langle x^*, x \rangle + (\bar{z}^* g)(x)\},$$

which leads to $0 \in \partial(f - \langle x^*, \cdot \rangle + (\bar{z}^* g) + \delta_S)(\bar{x})$, i.e.

$$x^* \in \partial(f + (\bar{z}^* g) + \delta_S)(\bar{x}) \subseteq \bigcup_{\substack{z^* \in C^* \\ (z^* g)(\bar{x}) = 0}} \partial(f + (z^* g) + \delta_S)(\bar{x}).$$

The inclusion "$\subseteq$" in (17.1) is fulfilled, too, and this concludes the proof. □

A similar statement to Theorem 17.1, involving the Fenchel–Lagrange dual problem of $\inf_{x\in\mathcal{A}}\{f(x)+\langle x^*,x\rangle\}$, where $x^*\in X^*$, is given in the following. The proof can be made in the lines of the one above and therefore we omit it. For more results of this kind we refer to [30].

Theorem 17.2. *Let be $\bar{x}\in \operatorname{dom} f\cap\mathcal{A}$. Then*

$$\partial(f+\delta_{\mathcal{A}})(\bar{x}) = \partial f(\bar{x}) + \bigcup_{\substack{z^*\in C^* \\ (z^*g)(\bar{x})=0}} \partial((z^*g)+\delta_S)(\bar{x}) \tag{17.3}$$

if and only if for each $x^*\in X^*$ for which the infimum over $\mathcal{A}$ of the function $f+\langle x^*,\cdot\rangle$ is attained at $\bar{x}$ one has

$$f(\bar{x})+\langle x^*,\bar{x}\rangle = \min_{x\in\mathcal{A}}\{f(x)+\langle x^*,x\rangle\} = \max_{y^*\in X^*,z^*\in C^*}\{-f^*(y^*)-(z^*g)_S^*(-y^*)\}. \tag{17.4}$$

When f is identical to 0, then (17.1) and (17.3) collapse into a sufficient condition which refers to the set of constraints $\mathcal{A}$. This kind of regularity conditions are called *constraint qualifications*. We say that the set $\mathcal{A}$ fulfills the *generalized Basic Constraint Qualification* $(gBCQ)$ *at* $x\in\mathcal{A}$ if

$$\partial\delta_{\mathcal{A}}(x) = \bigcup_{\substack{z^*\in C^* \\ (z^*g)(x)=0}} \partial((z^*g)+\delta_S)(x).$$

If the set $\mathcal{A}$ fulfills $(gBCQ)$ for all $x\in\mathcal{A}$ we say that $\mathcal{A}$ fulfills the *generalized Basic Constraint Qualification* $(gBCQ)$. The characterization of $(gBCQ)$ from below follows as a consequence of Theorem 17.1 (or Theorem 17.2).

Corollary 17.3. *The set $\mathcal{A}$ fulfills $(gBCQ)$ at $\bar{x}\in\mathcal{A}$ if and only if for all $x^*\in X^*$ such that $\langle x^*,\cdot\rangle$ attains its infimum over $\mathcal{A}$ at $\bar{x}$ one has* $\langle x^*,\bar{x}\rangle = \min_{x\in\mathcal{A}}\langle x^*,x\rangle = \max_{z^*\in C^*}\inf_{x\in S}\{\langle x^*,x\rangle+(z^*g)(x)\}$.

The next theorem characterizes by using $(gBCQ)$ the existence of strong duality for the convex optimization problem (P^C) and its three conjugate duals.

Theorem 17.4. *The set $\mathcal{A}$ fulfills $(gBCQ)$ at $\bar{x}\in\mathcal{A}$ if and only if for each proper, convex and lower semicontinuous function $f: X\to\overline{\mathbb{R}}$ such that* $\operatorname{dom} f\cap\mathcal{A}\neq\emptyset$ *which satisfies $(RC_5^{C_F})$ and attains its minimum over $\mathcal{A}$ at $\bar{x}$ one has that $v(P^C) = v(D^{C_L}) = v(D^{C_F}) = v(D^{C_{FL}})$ and the dual problems have optimal solutions.*

Proof. As the sufficiency follows obviously from the preceding corollary by taking f linear and continuous, we have only to show the necessity. To this aim, take f as requested in the hypothesis. We have $f(\bar{x}) = \inf_{x\in\mathcal{A}} f(x) = -(f+\delta_{\mathcal{A}})^*(0)$. As in the proof of Theorem 16.13, one gets that there exists some $\bar{y}^*\in\partial f(\bar{x})$ such that $-\bar{y}^*\in\partial\delta_{\mathcal{A}}(\bar{x})$. Since $(gBCQ)$ is fulfilled at $\bar{x}$, there is a $\bar{z}^*\in C^*$

such that $(\bar{z}^*g)(\bar{x}) = 0$ and $-\bar{y}^* \in \partial((\bar{z}^*g) + \delta_S)(\bar{x})$. Consequently, $((\bar{z}^*g) + \delta_S)(\bar{x}) + ((\bar{z}^*g) + \delta_S)^*(-\bar{y}^*) = \langle -\bar{y}^*, \bar{x}\rangle$ and this yields $f(\bar{x}) = -f^*(\bar{y}^*) - ((\bar{z}^*g) + \delta_S)^*(-\bar{y}^*) = 0$. Thus $v(P^C) = v(D^{C_{FL}})$ and $(\bar{y}^*, \bar{z}^*) \in X^* \times C^*$ is an optimal solution of the Fenchel–Lagrange dual problem. By Remark 3.7 the conclusion follows. □

By combining the last two results, we obtain the following characterization for $(gBCQ)$.

Theorem 17.5. *The following statements are equivalent:*

(i) $\mathcal{A}$ *fulfills the condition* $(gBCQ)$*;*
(ii) for each $x^* \in X^*$ *such that* $\langle x^*, \cdot\rangle$ *attains its minimum over* $\mathcal{A}$ *one has*

$$\min_{x\in\mathcal{A}}\langle x^*, x\rangle = \max_{z^*\in C^*}\inf_{x\in S}\{\langle x^*, x\rangle + (z^*g)(x)\};$$

(iii) for each proper, convex and lower semicontinuous function $f : X \to \overline{\mathbb{R}}$ *such that* $\operatorname{dom} f \cap \mathcal{A} \neq \emptyset$ *which satisfies* $(RC_5^{C_F})$ *and attains its minimum over* $\mathcal{A}$ *one has*

$$\min_{x\in\mathcal{A}} f(x) = \max_{z^*\in C^*}\inf_{x\in S}\{f(x) + (z^*g)(x)\}.$$

Remark 17.6. One can replace in Theorems 17.4 and 17.5 the regularity condition $(RC_5^{C_F})$ by any of the conditions $(RC_i^{C_F})$, $i \in \{1,2,3,4\}$, without altering the validity of the results. When g is continuous at some point of $\mathcal{A}$, $(gBCQ)$ turns at each $x \in \mathcal{A}$ into

$$\partial\delta_{\mathcal{A}}(x) = \bigcup_{\substack{z^*\in C^*\\ (z^*g)(x)=0}} \partial(z^*g)(x) + \delta_S(x).$$

Remark 17.7. As in Remark 16.12, let T be an arbitrary index set and consider a family of functions $g_t : X \to \mathbb{R}, t \in T$, which are convex and continuous at some point of $\{x \in S : g_t(x) \le 0, t \in T\}$. Take $C = \mathbb{R}^T_+$ and define $g : X \to \mathbb{R}^T$ as being $g(x) = (g_t(x))_{t\in T}$. In this setting, the condition $(gBCQ)$ at x becomes the so-called *locally Farkas–Minkowski condition* (LFM) *at* x (cf. [64, 69, 74])

$$\partial\delta_{\mathcal{A}}(x) = \operatorname{coneco}\left(\bigcup_{t\in T(x)}\partial g_t(x)\right) + \partial\delta_S(x),$$

where $T(x) = \{t \in T : g_t(x) = 0\}$. Indeed, using that $g_t, t \in T$, is continuous at some point in $\mathcal{A}$, from Theorems 5.5, 8.3 and 8.6 one has that $(gBCQ)$ at x turns into

$$\partial\delta_{\mathcal{A}}(x) = \bigcup_{\substack{z^*\in(\mathbb{R}^T_+)^*\\ (z^*g)(x)=0}} \partial\left(\sum_{t\in T} z_t^* g_t\right)(x) + \partial\delta_S(x)$$

$$= \bigcup_{\substack{z^* \in (\mathbb{R}^T_+)^* \\ (z^* g)(x)=0}} \left\{ \sum_{t \in T'} z_t^* \partial g_t(x) : T' \subseteq T, |T'| < +\infty, z_t^* > 0, g_t(x) = 0 \ \forall t \in T' \right\} \cup \{0\}$$

$$+ \partial \delta_S(x) = \operatorname{coneco} \left(\bigcup_{t \in T(x)} \partial g_t(x) \right) + \partial \delta_S(x).$$

Consequently, Theorem 17.5 yields a similar result like [74, Theorem 5.1], while it improves the conditions put on f in statement (iii).

We want to mention that recently an extended formulation of the locally Farkas–Minkowski condition, which can be also derived from the general subdifferential sum formulae given in Section 8, was employed in [65] in characterizing the subdifferentials of the infimal value functions of several classes of difference-convex (DC) and bilevel infinite and semi-infinite programs. Here techniques from the convex analysis join the ones coming from the variational analysis (see [99, 100]).

If T is a finite index set and $S = X$, $(gBCQ)$ is actually the condition (BCQ) considered in [123]. When $U = X$ and $x \in \operatorname{bd}(\mathcal{A})$, $(gBCQ)$ at x becomes the condition (BCQ) at x in [80]. Considering $A \in \mathbb{R}^{m \times n}$, $b \in \mathbb{R}^m$, the convex functions $c_j : \mathbb{R}^n \to \mathbb{R}$, $j = 1, \ldots, r$, and $\mathcal{A} = \{x \in \mathbb{R}^n : Ax = b, c_j(x) \leq 0, \ j = 1, \ldots, r\}$, $(gBCQ)$ becomes exactly the *basic constraint qualification* (BCQ) in its original formulation due to Hiriart-Urruty and Lemaréchal (cf. [79]).

Chapter V
Unconventional Fenchel Duality

18 Totally Fenchel Unstable Functions

In the fifth chapter of this work, we give some new insights into the classical Fenchel duality. The first two sections deal with the concept of *totally Fenchel unstable functions* introduced by Stephen Simons in [120], while in the remaining sections we turn our attention to the study of some "unconventional" regularity conditions for Fenchel duality expressed via the *quasi interior* and the *quasi-relative interior* of the domains and epigraphs of the functions involved.

Let X be a separated locally convex space, X^* its topological dual space and $f, g : X \to \overline{\mathbb{R}}$ two arbitrary proper functions. According to the terminology used in Section 5 (see also Definition 5.4), we say that f and g satisfy *stable Fenchel duality* if for all $x^* \in X^*$, there exists $y^* \in X^*$ such that $(f+g)^*(x^*) = f^*(x^* - y^*) + g^*(y^*)$. If this property holds for $x^* = 0$, then f, g satisfy the classical *Fenchel duality*. Due to Stephen Simons (see [120]), the pair f, g is said to be *totally Fenchel unstable* if f and g satisfy Fenchel duality, but

$$y^*, z^* \in X^* \text{ and } (f+g)^*(y^* + z^*) = f^*(y^*) + g^*(z^*) \Longrightarrow y^* + z^* = 0.$$

A geometric characterization of these notions in terms of the epigraphs of the conjugates of the functions involved is given below. As pointed out in the discussion made below Theorem 7.7, if f and g are proper functions such that $\operatorname{dom} f \cap \operatorname{dom} g \neq \emptyset$, then stable Fenchel duality is equivalent to the relation $\operatorname{epi}(f+g)^* = \operatorname{epi} f^* + \operatorname{epi} g^*$. As seen in Section 7, a necessary and sufficient condition for this equality whenever f and g are proper, convex and lower semicontinuous functions such that $\operatorname{dom} f \cap \operatorname{dom} g \neq \emptyset$ is to have that $\operatorname{epi} f^* + \operatorname{epi} g^*$ is closed in $(X^*, \omega(X^*, X)) \times \mathbb{R}$.

Whenever f and g are proper functions such that $\operatorname{dom} f \cap \operatorname{dom} g \neq \emptyset$ one can prove that Fenchel duality is equivalent to the relation

$$\operatorname{epi}(f+g)^* \cap (\{0\} \times \mathbb{R}) = (\operatorname{epi} f^* + \operatorname{epi} g^*) \cap (\{0\} \times \mathbb{R}).$$

R.I. Boţ, *Conjugate Duality in Convex Optimization*, Lecture Notes in Economics and Mathematical Systems 637, DOI 10.1007/978-3-642-04900-2_5,

The proof of this assertion can be gathered from the one of Proposition 18.1. Before coming to this result, we want to notice that in case f and g are proper, convex and lower semicontinuous functions such that $\operatorname{dom} f \cap \operatorname{dom} g \neq \emptyset$ Fenchel duality is nothing else than saying that $\operatorname{epi} f^* + \operatorname{epi} g^*$ is *closed regarding the set* $\{0\} \times \mathbb{R}$ in $(X^*, \omega(X^*, X)) \times \mathbb{R}$.

As for all $y^*, z^* \in X^*$ it holds

$$(f+g)^*(y^*+z^*) \leq f^*(y^*) + g^*(z^*), \tag{18.1}$$

then a pair f, g of proper functions is *totally Fenchel unstable* if and only if

$$\exists y^* \in X^* : \ (f+g)^*(0) = f^*(-y^*) + g^*(y^*) \tag{18.2}$$

and

$$\forall x^* \in X^* \setminus \{0\}, \forall z^* \in X^* : (f+g)^*(x^*) < f^*(x^*-z^*) + g^*(z^*). \tag{18.3}$$

This means that if the pair f, g is totally Fenchel unstable one must have that $\operatorname{dom} f \cap \operatorname{dom} g \neq \emptyset$. Indeed, if this is not the case, then $f+g$ is identical to $+\infty$ and, consequently, $(f+g)^*$ is identical to $-\infty$. By (18.2) there exists $y^* \in X^*$ such that $f^*(-y^*) + g^*(y^*) = -\infty$. But, f and g being proper, we get $f^*(-y^*) > -\infty$ and $g^*(y^*) > -\infty$, a contradiction.

Proposition 18.1. *Let* $f, g : X \to \overline{\mathbb{R}}$ *be proper functions such that* $\operatorname{dom} f \cap \operatorname{dom} g \neq \emptyset$. *Then, the pair* f, g *is totally Fenchel unstable if and only if*

$$\operatorname{epi}(f+g)^* \cap (\{0\} \times \mathbb{R}) = (\operatorname{epi} f^* + \operatorname{epi} g^*) \cap (\{0\} \times \mathbb{R}) \tag{18.4}$$

and there is no $x^* \in X^* \setminus \{0\}$ *such that*

$$\operatorname{epi}(f+g)^* \cap (\{x^*\} \times \mathbb{R}) = (\operatorname{epi} f^* + \operatorname{epi} g^*) \cap (\{x^*\} \times \mathbb{R}). \tag{18.5}$$

Proof. We notice first that we always have $\operatorname{epi}(f+g)^* \supseteq \operatorname{epi} f^* + \operatorname{epi} g^*$. As $\operatorname{dom} f \cap \operatorname{dom} g \neq \emptyset$, $(f+g)^*$ never attains $-\infty$.

"$\Rightarrow$" In case $(f+g)^*(0) = +\infty$, the set $\operatorname{epi}(f+g)^* \cap (\{0\} \times \mathbb{R})$ is empty and (18.4) follows automatically. In case $(f+g)^*(0) \in \mathbb{R}$, take an arbitrary element $r \in \mathbb{R}$ fulfilling $(f+g)^*(0) \leq r$. By (18.2) there exists $y^* \in X^*$ such that $f^*(-y^*) + g^*(y^*) \leq r$ and so

$$(0, r) = (-y^*, f^*(-y^*)) + (y^*, r - f^*(-y^*)) \in (\operatorname{epi} f^* + \operatorname{epi} g^*) \cap (\{0\} \times \mathbb{R}).$$

Also in this case (18.4) follows.

Assume now that for $x^* \in X^* \setminus \{0\}$ relation (18.5) is fulfilled. As (18.3) implies $(f+g)^*(x^*) < +\infty$, we have $(f+g)^*(x^*) \in \mathbb{R}$. In this case $(x^*, (f+g)^*(x^*)) \in \operatorname{epi}(f+g)^* \cap (\{x^*\} \times \mathbb{R})$ and so $(x^*, (f+g)^*(x^*)) \in \operatorname{epi} f^* + \operatorname{epi} g^*$. Thus there exist $(y^*, s) \in \operatorname{epi} f^*$ and $(z^*, t) \in \operatorname{epi} g^*$ such that $y^* + z^* = x^*$ and

$s+t = (f+g)^*(x^*)$. This means that $f^*(y^*) + g^*(z^*) \leq (f+g)^*(y^*+z^*)$, which contradicts (18.3).

"$\Leftarrow$" We prove first that Fenchel duality holds. If $(f+g)^*(0) = +\infty$ this follows automatically from (18.1). If $(f+g)^*(0) \in \mathbb{R}$, then $(0,(f+g)^*(0)) \in \operatorname{epi} f^* + \operatorname{epi} g^*$ and so there exist $(-y^*, s) \in \operatorname{epi} f^*$ and $(y^*, t) \in \operatorname{epi} g^*$ such that $s+t = (f+g)^*(0)$. Thus $f^*(-y^*) + g^*(y^*) \leq (f+g)^*(0)$ and the conclusion follows.

Further, assume that there exist $y^*, z^* \in X^*$ such that $y^* + z^* \neq 0$ and $(f+g)^*(y^*+z^*) = f^*(y^*) + g^*(z^*)$. As (18.5) does not hold with equality we get $(f+g)^*(y^*+z^*) \in \mathbb{R}$. For all $r \in \mathbb{R}$ such that $(f+g)^*(y^*+z^*) \leq r$ it holds $(y^*+z^*, r) \in (\operatorname{epi} f^* + \operatorname{epi} g^*) \cap (\{y^*+z^*\} \times \mathbb{R})$. This implies that (18.5) is satisfied for $x^* = y^* + z^* \neq 0$, a contradiction. □

Obviously, stable Fenchel duality implies Fenchel duality, but the converse is not true (see for instance Example 15.3 but also the example in [40, pp. 2798–2799] and Example 11.1 in [120]). Nevertheless, each of these examples fails when one tries to verify total Fenchel unstability. In the infinite dimensional setting the problem of providing a pair of proper and convex totally Fenchel unstable functions received an answer, due to the existence of *extreme points* which are not *support points* of certain convex sets. Recall that if C is a convex subset of X, then $x \in C$ is a *support point* of C if there exists $x^* \in X^*$, $x^* \neq 0$, such that $\langle x^*, x\rangle = \sup_{c\in C} \langle x^*, c\rangle$.

We give below an example proposed in [120] of a pair f, g which is totally Fenchel unstable. To this aim, we suppose that X is a nonzero real Banach space and that its topological dual X^* is endowed with the norm topology.

Example 18.2. (cf. [120, Example 11.3]) Let C be a nonempty, bounded, convex and closed subset of X such that there exists an extreme point x_0 of C which is not a support point of C. Take $f := \delta_{x_0 - C}$ and $g := \delta_{C - x_0}$. The fact that x_0 is an extreme point of C implies that $f + g = \delta_{\{0\}}$. The conjugates of the functions f and g are

$$f^*(y^*) = \langle y^*, x_0\rangle - \inf_{c\in C} \langle y^*, c\rangle \geq 0 \ \forall y^* \in X^*$$

and

$$g^*(z^*) = \sup_{c\in C} \langle z^*, c\rangle - \langle z^*, x_0\rangle \geq 0 \ \forall z^* \in X^*.$$

Since $(f+g)^*(0) = f^*(0) + g^*(0)$, f and g satisfy Fenchel duality. For $x^* \in X^*$ we consider $y^*, z^* \in X^*$ such that $y^* + z^* = x^*$ and $f^*(y^*) + g^*(z^*) = (f+g)^*(x^*) = 0$. It follows that $f^*(y^*) = g^*(z^*) = 0$. Consequently, $\inf_{c\in C} \langle y^*, c\rangle = \langle y^*, x_0\rangle$ and $\sup_{c\in C} \langle z^*, c\rangle = \langle z^*, x_0\rangle$. As x_0 is not a support point of C we must have $y^* = z^* = 0$. Thus $x^* = 0$ and the pair f, g fails to be totally Fenchel unstable.

An example of a set C and a point x_0 with the above mentioned properties was also given in [120], following an idea due to Jonathan Borwein. For $X = \ell^2$, $1 < p < 2$, $C = \{x \in \ell^2 : \|x\|_p \leq 1\}$ one has that x is an extreme point of C if and only if $\|x\|_p = 1$. For $x \in C$ with $\|x\|_p = 1$ it yields that x is a support point of C

if and only if $x \in \ell^{2(p-1)}$ (cf. [120]). As $2(p-1) < p$ there exist extreme points of C which are not support points.

Regarding the functions defined in the above example, Stephen Simons asked whether the following representation of the Minkowski sum of the sets epi f^* and epi g^* is true:

$$\operatorname{epi} f^* + \operatorname{epi} g^* = \{(0,0)\} \cup \big(X^* \times (0,+\infty)\big). \tag{18.6}$$

The inclusion "$\subseteq$" in (18.6) is obvious and, since $(0,0) = (0,0) + (0,0) \in \operatorname{epi} f^* + \operatorname{epi} g^*$, relation (18.6) is equivalent to

$$\operatorname{epi} f^* + \operatorname{epi} g^* \supset X^* \times (0,+\infty). \tag{18.7}$$

Let us mention that for the implication (18.7)$\Rightarrow$(18.6) the assumption that x_0 is not a support point of C is decisive. One should also notice that the boundedness of the set C guarantees that $\operatorname{dom} f^* = \operatorname{dom} g^* = X^*$, thus f^* and g^* are continuous functions (see [127, Theorem 2.2.9]).

In case X is reflexive Simon's question gets a positive answer. Although the proof is given in [120, Example 11.3], we give here the details for the reader's convenience. Let $y^* \in X^*$ be arbitrary. Consider the functions $h : X^* \to \mathbb{R}$ and $k : X^* \to \mathbb{R}$ defined by $h(z^*) := f^*(z^*)$ and $k(z^*) := g^*(y^* - z^*)$ for all $z^* \in X^*$. Since h and k are continuous, it follows that h and k satisfy Fenchel duality (see Theorem 2.2). This and the reflexivity of the space X gives

$$-\inf_{z^* \in X^*} \{h(z^*) + k(z^*)\} = (h+k)^*(0) = \min_{z \in X}\{h^*(z) + k^*(-z)\}.$$

A simple computation shows that $h^*(z) = f(z)$ and $k^*(-z) = g(z) - \langle y^*, z\rangle$ for all $z \in X$. Hence

$$-\inf_{z^* \in X^*} \{h(z^*) + k(z^*)\} = \min_{z \in X}\{f(z) + g(z) - \langle y^*, z\rangle\} = \min_{z \in X}\{\delta_{\{0\}}(z) - \langle y^*, z\rangle\} = 0,$$

consequently, for all $\varepsilon > 0$, there exists $z^* \in X^*$ such that $h(z^*) + k(z^*) \le \varepsilon$, that is $f^*(z^*) + g^*(y^* - z^*) \le \varepsilon$. This means exactly that $(y^*, \varepsilon) \in \operatorname{epi} f^* + \operatorname{epi} g^*$, hence the proof of (18.7) is complete.

Remark 18.3. With respect to the proof given above, one can notice that relation (18.6) is fulfilled if and only if for all $y^* \in X^*$ and for all $\varepsilon > 0$ there exists $z^* \in X^*$ such that $f^*(z^*) + g^*(y^* - z^*) \le \varepsilon$. This is further equivalent to the statement that there exists $z^* \in X^*$ such that for all $x, y \in X$, $f(x) + g(y) - \langle z^*, x - y\rangle \ge \langle y^*, y\rangle - \varepsilon$. Using the Hahn–Banach–Lagrange theorem (see [120, Theorem 1.11]), this is equivalent to the following: there exists $M \ge 0$ such that for all $x, y \in X$, $f(x) + g(y) + M\|x - y\| \ge \langle y^*, y\rangle - \varepsilon$, that is to say there exists $M \ge 0$ such that for all $u, v \in C$, $M\|u + v - 2x_0\| \ge \langle y^*, v - x_0\rangle - \varepsilon$.

Following this observation Stephen Simons proposed the following problem (see [120, Problem 11.5]).

Problem 18.4. Let C be a nonempty, bounded, convex and closed subset of a nonreflexive Banach space X, x_0 be an extreme point of C, $y^* \in X^*$ and $\varepsilon > 0$. Then does there always exist $M \geq 0$ such that for all $u, v \in C$, $M\|u + v - 2x_0\| \geq \langle v - x_0, y^*\rangle - \varepsilon$? If the answer to this question is positive, then

$$\operatorname{epi} \delta^*_{x_0 - C} + \operatorname{epi} \delta^*_{C - x_0} \supset X^* \times (0, +\infty).$$

In the following, we give an answer to Problem 18.4 and show that in the nonreflexive case the answer depends on whether x_0 is a weak*-extreme point of C or not. We recall that x_0 is a *weak*-extreme point* (see [92]) of the nonempty, bounded, convex and closed set $C \subseteq X$ if $\widehat{x_0}$ is an extreme point of $\operatorname{cl}_{\omega^*}(\widehat{C})$, where the closure is taken with respect to the weak* topology $\omega(X^{**}, X^*)$. One can easily show that if x_0 is a weak*-extreme point of C, then x_0 is an extreme point of C.

Remark 18.5. The history of the notion of weak*-extreme point goes back to the paper of Phelps [109], where the author asked the following: must the image $\widehat{x}$ of an extreme point of $x \in B_X$ (the unit ball of X) be an extreme point of $B_{X^{**}}$ (the unit ball of the bidual)? We recall that by the Goldstine theorem (see pp. 126–127 in [77]), the closure of $\widehat{B_X}$ in the weak* topology $\omega(X^{**}, X^*)$ is $B_{X^{**}}$, hence the generalization to a nonempty, bounded, convex and closed set is natural. In the spaces $C(X)$ and $L^p(1 \leq p \leq \infty)$ all the extreme points of the corresponding unit balls are weak*-extreme points (see [101]). The first example of a Banach space the unit ball of which contains elements which are not weak*-extreme was suggested by K. de Leeuw and proved by Y. Katznelson (see the note added at the end of [109]). For the general case when C is a bounded, convex and closed set we refer to [43] and [92] for more on this subject. We recall from [43] the following result: a Banach space X has the Radon–Nikodým property if and only if every bounded, convex and closed subset C of X has a weak*-extreme point. Of course, in a Radon–Nikodým space it is possible that some of the extreme points are not weak*-extreme points.

The answer to Problem 18.4 will be the consequence of some general results that we prove in the following. Consider the proper and convex functions $f, g : X \to \overline{\mathbb{R}}$ with the properties $\operatorname{dom} f \cap \operatorname{dom} g \neq \emptyset$, $\operatorname{cl}_{\omega^*} \widehat{f}$ and $\operatorname{cl}_{\omega^*} \widehat{g}$ are proper, $f^{**}(0) + g^{**}(0) \geq 0$ and $\operatorname{dom} f^* + \operatorname{dom} g^* = X^*$. Here $\widehat{f} : X^{**} \to \overline{\mathbb{R}}$ is the function we attached to f in Lemma 10.1, defined by $\widehat{f}(x^{**}) = f(x)$, if $x^{**} = \widehat{x} \in \widehat{X}$ and $\widehat{f}(x^{**}) = +\infty$, otherwise, while $\operatorname{cl}_{\omega^*} \widehat{f}$ is the lower semicontinuous hull of $\widehat{f}$ in the topology $\omega(X^{**}, X^*)$. The same applies for $\widehat{g}$ and $\operatorname{cl}_{\omega^*} \widehat{g}$, respectively. Define the function $P : X^* \to \overline{\mathbb{R}}$, $P(z^*) = (f^{**} + g^{**})^*(z^*)$.

Let $y^* \in X^*$ be fixed. Consider also the functions $h : X^* \to \overline{\mathbb{R}}$ and $k : X^* \to \overline{\mathbb{R}}$ defined by $h(z^*) := f^*(z^*)$ and $k(z^*) := g^*(y^* - z^*)$ for all $z^* \in X^*$.

Lemma 18.6. *We have*

(i) $\inf_{z^* \in X^*}\{h(z^*) + k(z^*)\} = P(y^*)$*;*
(ii) If $\lambda \in \mathbb{R}$ *then*

$$(y^*, \lambda) \in \operatorname{epi} f^* + \operatorname{epi} g^* \Leftrightarrow \textit{there exists } z^* \in X^* \textit{ such that } (h + k)(z^*) \le \lambda.$$

Proof. (i) Since $\operatorname{dom} h = \operatorname{dom} f^*$ and $\operatorname{dom} k = y^* - \operatorname{dom} g^*$, we get $\operatorname{dom} h - \operatorname{dom} k = -y^* + \operatorname{dom} f^* + \operatorname{dom} g^* = X^*$. It follows that for h and k the regularity condition (RC_2^{id}) is fulfilled and thus, by Theorem 2.2, we obtain

$$\inf_{z^* \in X^*}\{h(z^*) + k(z^*)\} = \max_{z^{**} \in X^{**}}\{-h^*(z^{**}) - k^*(-z^{**})\}.$$

As $h^*(z^{**}) = f^{**}(z^{**})$ and $k^*(-z^{**}) = g^{**}(z^{**}) - \langle z^{**}, y^* \rangle$ for all $z^{**} \in X^{**}$ the conclusion follows easily.

(ii) This part is immediate from the definition of the functions h and k. □

Lemma 18.7. *Let* $(y^*, \lambda) \in X^* \times \mathbb{R}$*. Then:*

$$\lambda > P(y^*) \Rightarrow (y^*, \lambda) \in \operatorname{epi} f^* + \operatorname{epi} g^* \Rightarrow \lambda \ge P(y^*).$$

Proof. If $\lambda > P(y^*)$, then Lemma 18.6(i) gives an $z^* \in X^*$ such that $(h+k)(z^*) < \lambda$ and so Lemma 18.6(ii) implies that $(y^*, \lambda) \in \operatorname{epi} f^* + \operatorname{epi} g^*$.

On the other hand, if $(y^*, \lambda) \in \operatorname{epi} f^* + \operatorname{epi} g^*$, from Lemma 18.6(ii), there exists $z^* \in X^*$ such that $(h + k)(z^*) \le \lambda$. Hence $\inf_{x^* \in X^*}(h + k)(x^*) \le \lambda$ and so, from Lemma 18.6(i), we obtain $\lambda \ge P(y^*)$. □

We come now to the result which leads to the answer to Problem 18.4 (see also [16]).

Theorem 18.8. *We have* $X^* \times (0, +\infty) \subset \operatorname{epi} f^* + \operatorname{epi} g^*$ *if and only if* $\operatorname{dom}(\operatorname{cl}_{\omega^*} \widehat{f}) \cap \operatorname{dom}(\operatorname{cl}_{\omega^*} \widehat{g}) = \{0\}$.

Proof. By Lemma 10.1 we have for all $y^* \in X^*$ that

$$P(y^*) = \sup_{z^{**} \in \operatorname{dom}(\operatorname{cl}_{\omega^*} \widehat{f}) \cap \operatorname{dom}(\operatorname{cl}_{\omega^*} \widehat{g})} \left\{\langle z^{**}, y^* \rangle - \operatorname{cl}_{\omega^*} \widehat{f}(z^{**}) - \operatorname{cl}_{\omega^*} \widehat{g}(z^{**})\right\}.$$

Let us suppose first that $X^* \times (0, +\infty) \subset \operatorname{epi} f^* + \operatorname{epi} g^*$. Thus, for all $y^* \in X^*$ and $\lambda > 0$, $(y^*, \lambda) \in \operatorname{epi} f^* + \operatorname{epi} g^*$ and so, from Lemma 18.7, we get $P(y^*) \le \lambda$. We obtain $P(y^*) \le 0$ for all $y^* \in X^*$, that is $\langle z^{**}, y^* \rangle - \operatorname{cl}_{\omega^*} \widehat{f}(z^{**}) - \operatorname{cl}_{\omega^*} \widehat{g}(z^{**}) \le 0$ for all $y^* \in X^*$ and $z^{**} \in \operatorname{dom}(\operatorname{cl}_{\omega^*} \widehat{f}) \cap \operatorname{dom}(\operatorname{cl}_{\omega^*} \widehat{g})$, from which follows that $\operatorname{dom}(\operatorname{cl}_{\omega^*} \widehat{f}) \cap \operatorname{dom}(\operatorname{cl}_{\omega^*} \widehat{g}) = \{0\}$.

For the reverse implication, one can notice that if $\operatorname{dom}(\operatorname{cl}_{\omega^*} \widehat{f}) \cap \operatorname{dom}(\operatorname{cl}_{\omega^*} \widehat{g}) = \{0\}$ one has $P(y^*) = -\operatorname{cl}_{\omega^*} \widehat{f}(0) - \operatorname{cl}_{\omega^*} \widehat{g}(0) = -f^{**}(0) - g^{**}(0) \le 0$ for all $y^* \in X^*$. From Lemma 18.7, for all $y^* \in X^*$ and $\lambda > 0$, $(y^*, \lambda) \in \operatorname{epi} f^* + \operatorname{epi} g^*$, hence $X^* \times (0, +\infty) \subset \operatorname{epi} f^* + \operatorname{epi} g^*$. □

Consider now the following particular setting: C is a nonempty, bounded and convex subset of the Banach space X, $x_0 \in C$, $f := \delta_{x_0 - C}$ and $g := \delta_{C - x_0}$. We have $f^* = \delta^*_{x_0 - C}$, $g^* = \delta^*_{C - x_0}$, $\operatorname{dom} f^* = \operatorname{dom} g^* = X^*$, $\widehat{f} = \delta_{\widehat{x}_0 - \widehat{C}}$, $\operatorname{cl}_{\omega^*} \widehat{f} = \delta_{\widehat{x}_0 - \operatorname{cl}_{\omega^*}(\widehat{C})}$, thus, in view of Corollary 10.2, $f^{**} = \delta_{\widehat{x}_0 - \operatorname{cl}_{\omega^*}(\widehat{C})}$. Further, $g^{**} = \delta_{\operatorname{cl}_{\omega^*}(\widehat{C}) - \widehat{x}_0}$. Applying Theorem 18.8 to this particular instance (the hypotheses regarding the functions f and g are obviously fulfilled), we obtain the following result.

Corollary 18.9. *We have* $X^* \times (0, +\infty) \subset \operatorname{epi} \delta^*_{x_0 - C} + \operatorname{epi} \delta^*_{C - x_0}$ *if and only if* x_0 *is a weak***-extreme point of* C.

Remark 18.10. Corollary 18.9 shows that the answer to Problem 18.4 is affirmative if and only if x_0 is a weak*-extreme point of C. The closedness of the set C, requested in [120], is not necessary for this result.

We conclude the section by providing via a minmax result an alternative proof of Corollary 18.9, hence an alternative solution to Problem 18.4 (see [16]).

Alternative solution to Problem 18.4. Let $y^* \in X^*$ and $\varepsilon > 0$ be arbitrary. In view of Remark 18.3, the condition $(y^*, \varepsilon) \in \operatorname{epi} f^* + \operatorname{epi} g^*$ is equivalent to the statement that there exists $z^* \in X^*$ such that for all $x, y \in X$ $f(x) + g(y) - \langle z^*, x - y \rangle \geq \langle y^*, y \rangle - \varepsilon$, which is nothing else than there exists $z^* \in X^*$ such that for all $u, v \in C$, $\langle u + v - 2x_0, z^* \rangle + \langle y^*, x_0 - v \rangle \geq -\varepsilon$. Hence the inclusion $X^* \times (0, +\infty) \subset \operatorname{epi} f^* + \operatorname{epi} g^*$ is fulfilled if and only if

$$\sup_{z^* \in X^*} \inf_{(u,v) \in C \times C} \{\langle z^*, u + v - 2x_0 \rangle + \langle y^*, x_0 - v \rangle\} \geq 0 \ \forall y^* \in X^*. \tag{18.8}$$

Suppose first that x_0 is a weak*-extreme point of C. Take $y^* \in X^*$. For $z^* \in X^*$ we have

$$\inf_{(u,v) \in C \times C} \{\langle z^*, u + v - 2x_0 \rangle + \langle y^*, x_0 - v \rangle\} = \inf_{(u,v) \in \widehat{C} \times \widehat{C}} \{\langle u + v - 2\widehat{x}_0, z^* \rangle + \langle \widehat{x}_0 - v, y^* \rangle\}$$

$$= \inf_{(u,v) \in \operatorname{cl}_{\omega^*}(\widehat{C}) \times \operatorname{cl}_{\omega^*}(\widehat{C})} \{\langle u + v - 2\widehat{x}_0, z^* \rangle + \langle \widehat{x}_0 - v, y^* \rangle\},$$

where the first equality follows by the definition of the canonical embedding and the second one from the continuity (in the weak* topology $\omega(X^{**}, X^*)$) of the functions $\langle \cdot, x^* \rangle : X^{**} \to \mathbb{R}$ for all $x^* \in X^*$. The set C being bounded, by the Banach–Alaoglu Theorem, we conclude that $\operatorname{cl}_{\omega^*}(\widehat{C})$ is weak*-compact. We apply a minmax theorem (see, for example, [119, Theorem 3.1]) and obtain that

$$\sup_{z^* \in X^*} \inf_{(u,v) \in C \times C} \{\langle z^*, u + v - 2x_0 \rangle + \langle y^*, x_0 - v \rangle\}$$

$$= \sup_{z^* \in X^*} \inf_{(u,v) \in \operatorname{cl}_{\omega^*}(\widehat{C}) \times \operatorname{cl}_{\omega^*}(\widehat{C})} \{\langle u + v - 2\widehat{x}_0, z^* \rangle + \langle \widehat{x}_0 - v, y^* \rangle\}$$

$$= \inf_{(u,v)\in \mathrm{cl}_{\omega^*}(\widehat{C})\times \mathrm{cl}_{\omega^*}(\widehat{C})} \sup_{z^*\in X^*} \{\langle u+v-2\widehat{x}_0, z^*\rangle + \langle \widehat{x}_0 - v, y^*\rangle\}$$

$$= \inf_{\substack{(u,v)\in \mathrm{cl}_{\omega^*}(\widehat{C})\times \mathrm{cl}_{\omega^*}(\widehat{C})\\ u+v=2\widehat{x}_0}} \langle \widehat{x}_0 - v, y^*\rangle. \tag{18.9}$$

As x_0 is a weak*-extreme point of C, we get that $\{(u,v) \in \mathrm{cl}_{\omega^*}(\widehat{C}) \times \mathrm{cl}_{\omega^*}(\widehat{C}) : u+v = 2\widehat{x}_0\} = \{(\widehat{x}_0, \widehat{x}_0)\}$, hence

$$\inf_{\substack{(u,v)\in \mathrm{cl}_{\omega^*}(\widehat{C})\times \mathrm{cl}_{\omega^*}(\widehat{C})\\ u+v=2\widehat{x}_0}} \langle \widehat{x}_0 - v, y^*\rangle = 0.$$

Thus relation (18.8) is fulfilled, implying $X^* \times (0,+\infty) \subset \operatorname{epi} f^* + \operatorname{epi} g^*$.

For the reverse implication, we assume that $X^* \times (0,+\infty) \subset \operatorname{epi} f^* + \operatorname{epi} g^*$ and that x_0 is not a weak*-extreme point of C. Then, there exist $u_0, v_0 \in \mathrm{cl}_{\omega^*}(\widehat{C}) \times \mathrm{cl}_{\omega^*}(\widehat{C})$ with $u_0 + v_0 = 2\widehat{x}_0$ such that $u_0 \neq \widehat{x}_0$ and $v_0 \neq \widehat{x}_0$. We choose $y_0^* \in X^*$ such that $\langle \widehat{x}_0 - v_0, y_0^*\rangle < 0$. So there exists $\varepsilon_0 > 0$ such that $\langle \widehat{x}_0 - v_0, y_0^*\rangle < -\varepsilon_0$, hence

$$\inf_{\substack{(u,v)\in \mathrm{cl}_{\omega^*}(\widehat{C})\times \mathrm{cl}_{\omega^*}(\widehat{C})\\ u+v=2\widehat{x}_0}} \langle \widehat{x}_0 - v_0, y_0^*\rangle < -\varepsilon_0.$$

From (18.9) we get

$$\sup_{z^*\in X^*} \inf_{(u,v)\in C\times C} \{\langle z^*, u+v-2x_0\rangle + \langle y_0^*, x_0 - v\rangle\} < -\varepsilon_0 < 0,$$

which contradicts (18.8), hence the proof is complete. □

19 Totally Fenchel Unstable Functions in Finite Dimensional Spaces

We pursue the investigations on the concept of totally Fenchel unstable function and provide in this section an answer to another open problem posed by Stephen Simons (see [120, Problem 11.6]), which has the following formulation.

Problem 19.1. Do there exist a nonzero finite dimensional Banach space and a pair of extended real-valued, proper and convex functions which is totally Fenchel unstable?

As proved in the following, this question receives a negative answer and this can be interpreted as follows: *if two proper and convex functions $f, g : \mathbb{R}^n \to \overline{\mathbb{R}}$ satisfy Fenchel duality, then there exists at least one element $x^* \in \mathbb{R}^n \setminus \{0\}$ such that $f - \langle x^*, \cdot\rangle$ and g (or f and $g - \langle x^*, \cdot\rangle$) satisfy Fenchel duality, too.* We prove first a preliminary result.

Proposition 19.2. *Let* $f, g : \mathbb{R}^n \to \overline{\mathbb{R}}$ *be proper and convex functions such that* $\operatorname{int}(\operatorname{dom}(\operatorname{cl} f) \cap \operatorname{dom}(\operatorname{cl} g)) \neq \emptyset$. *Then the pair* f, g *satisfies stable Fenchel duality.*

Proof. Let $x' \in \operatorname{int}(\operatorname{dom}(\operatorname{cl} f) \cap \operatorname{dom}(\operatorname{cl} g)) \subseteq \operatorname{int}(\operatorname{dom}(\operatorname{cl} f)) \cap \operatorname{int}(\operatorname{dom}(\operatorname{cl} g))$. It holds

$$\operatorname{int}(\operatorname{dom}(\operatorname{cl} f)) = \operatorname{ri}(\operatorname{cl}(\operatorname{dom}(\operatorname{cl} f))) = \operatorname{ri}(\operatorname{cl}(\operatorname{dom} f)) = \operatorname{ri}(\operatorname{dom} f)$$

and the same applies for g. This means that $x' \in \operatorname{ri}(\operatorname{dom} f) \cap \operatorname{ri}(\operatorname{dom} g)$. The regularity condition (RC^{id}_{fin}) is fulfilled and, by Theorem 5.5, it follows that for all $x^* \in \mathbb{R}^n$ there exists some $y^* \in \mathbb{R}^n$ such that $(f+g)^*(x^*) = f^*(x^* - y^*) + g^*(y^*)$. This provides the conclusion. □

The announced result follows (see [35, Theorem 4]).

Theorem 19.3. *There are no proper convex functions* $f, g : \mathbb{R}^n \to \overline{\mathbb{R}}$ *such that the pair* f, g *is totally Fenchel unstable.*

Proof. We assume the contrary, namely that there exist $f, g : \mathbb{R}^n \to \overline{\mathbb{R}}$ proper and convex functions such that the pair f, g is totally Fenchel unstable. By (18.3), it follows that $(f+g)^*(x^*) < +\infty$ for all $x^* \in \mathbb{R}^n \setminus \{0\}$. The function $(f+g)^*$ being convex, we get $(f+g)^*(0) < +\infty$. As noticed in the previous section, we must have $\operatorname{dom} f \cap \operatorname{dom} g \neq \emptyset$, hence $(f+g)^*(0) > -\infty$.

This allows to choose some $\bar{x} \in \operatorname{dom} f \cap \operatorname{dom} g \subseteq \operatorname{dom}(\operatorname{cl} f) \cap \operatorname{dom}(\operatorname{cl} g)$ and consider $L = \operatorname{aff}(\operatorname{dom}(\operatorname{cl} f) \cap \operatorname{dom}(\operatorname{cl} g) - \bar{x}) = \operatorname{lin}(\operatorname{dom}(\operatorname{cl} f) \cap \operatorname{dom}(\operatorname{cl} g) - \bar{x})$. By Proposition 19.2, it yields $\operatorname{int}(\operatorname{dom}(\operatorname{cl} f) \cap \operatorname{dom}(\operatorname{cl} g)) = \emptyset$ and therefore the dimension of L is strictly less than n. This means that the orthogonal space to L, $L^\perp$ is nonzero. Obviously, we have

$$\operatorname{dom} f \cap \operatorname{dom} g \subseteq \operatorname{dom}(\operatorname{cl} f) \cap \operatorname{dom}(\operatorname{cl} g) \subseteq L + \bar{x} \tag{19.1}$$

By Theorem 7.6(ii), using that $f^* = (\operatorname{cl} f)^*$ and $g^* = (\operatorname{cl} g)^*$, it yields

$$\operatorname{epi}(\operatorname{cl} f + \operatorname{cl} g)^* = \operatorname{cl}(\operatorname{epi} f^* + \operatorname{epi} g^*). \tag{19.2}$$

It follows

$$\operatorname{epi}(f+g)^* \supseteq \operatorname{epi}(\operatorname{cl} f + \operatorname{cl} g)^* \supseteq \operatorname{epi} f^* + \operatorname{epi} g^*.$$

Since the pair f, g is totally Fenchel unstable, by Proposition 18.1, one has that

$$\operatorname{epi}(f+g)^* \cap (\{0\} \times \mathbb{R}) = \operatorname{epi}(\operatorname{cl} f + \operatorname{cl} g)^* \cap (\{0\} \times \mathbb{R}) = (\operatorname{epi} f^* + \operatorname{epi} g^*) \cap (\{0\} \times \mathbb{R})$$

and so $(f+g)^*(0) = (\operatorname{cl} f + \operatorname{cl} g)^*(0)$. Taking an element $x^* \in L^\perp \setminus \{0\}$ we obtain, by (19.1),

$$(f+g)^*(x^*) = \sup_{x \in \mathbb{R}^n} \{\langle x^*, x \rangle - f(x) - g(x)\} = \sup_{x \in L + \bar{x}} \{\langle x^*, x \rangle - f(x) - g(x)\}$$

$$= \langle x^*, \bar{x}\rangle + (f+g)^*(0) = \langle x^*, \bar{x}\rangle + (\operatorname{cl} f + \operatorname{cl} g)^*(0) = (\operatorname{cl} f + \operatorname{cl} g)^*(x^*). \tag{19.3}$$

We distinguish two cases:

(a) If $\operatorname{epi} f^* + \operatorname{epi} g^*$ is closed, then we obtain from (19.2) and (19.3) that $(x^*, (f+g)^*(x^*)) \in \operatorname{epi}(\operatorname{cl} f + \operatorname{cl} g)^* = \operatorname{epi} f^* + \operatorname{epi} g^*$ and so there exist $(y^*, s) \in \operatorname{epi} f^*$ and $(z^*, t) \in \operatorname{epi} g^*$ such that $y^* + z^* = x^* \neq 0$ and $s + t = (f+g)^*(x^*)$. This means that $f^*(y^*) + g^*(z^*) \leq (f+g)^*(y^* + z^*)$. As $y^* + z^* = x^* \neq 0$ this contradicts (18.3).

(b) Otherwise, if $\operatorname{epi} f^* + \operatorname{epi} g^*$ is not closed, by [113, Corollary 9.1.2] there exists a direction of recession of $\operatorname{epi} f^*$ whose opposite direction is a direction of recession of $\operatorname{epi} g^*$. This can be expressed as

$$\exists (x^*, r) \neq 0 \text{ such that } (x^*, r) \in 0^+ \operatorname{epi} f^* \text{ and } (-x^*, -r) \in 0^+ \operatorname{epi} g^*,$$

where by $0^+ \operatorname{epi} f^*$ and $0^+ \operatorname{epi} g^*$ we denote the recession cones of $\operatorname{epi} f^*$ and $\operatorname{epi} g^*$, respectively. Without loss of generality we assume that $r \geq 0$. It follows $x^* \neq 0$, because otherwise we would have $(0, -r) \in 0^+ \operatorname{epi} g^*$ with $r > 0$. But g is proper and so g^* never attains $-\infty$.

Choose some $y^* \in \mathbb{R}^n$ according to (18.2), namely such that $(f+g)^*(0) = f^*(-y^*) + g^*(y^*)$. Since $(f+g)^*(0), f^*(-y^*), g^*(y^*) \in \mathbb{R}$ and as $\operatorname{epi} f^*$ and $\operatorname{epi} g^*$ are nonempty convex sets, by [113, Theorem 8.1] it holds

$$(-y^*, f^*(-y^*)) + \lambda(x^*, r) \in \operatorname{epi} f^* \ \forall \lambda \geq 0$$

and

$$(y^*, g^*(y^*)) - \mu(x^*, r) \in \operatorname{epi} g^* \ \forall \mu \geq 0.$$

From the relations above we get

$$(0, (f+g)^*(0)) + \gamma(x^*, r) \in \operatorname{epi} f^* + \operatorname{epi} g^* \ \forall \gamma \in \mathbb{R}. \tag{19.4}$$

Let $\gamma = 1$ in (19.4). Thus there exist $(u^*, s) \in \operatorname{epi} f^*$ and $(v^*, t) \in \operatorname{epi} g^*$ such that $u^* + v^* = x^*$ and $s + t = (f+g)^*(0) + r$. It follows

$$(f+g)^*(x^*) \leq f^*(u^*) + g^*(v^*) \leq s + t = (f+g)^*(0) + r. \tag{19.5}$$

Setting $\gamma = -1$ in (19.4) we obtain analogously

$$(f+g)^*(-x^*) \leq (f+g)^*(0) - r. \tag{19.6}$$

Both inequalities in (19.5) and (19.6) must hold as equalities. Indeed, supposing that this is not the case, after adding them, we get a contradiction to the fact that $(f+g)^*$ is convex. Hence $(f+g)^*(u^* + v^*) = f^*(u^*) + g^*(v^*)$. But this contradicts (18.3) because of $u^* + v^* = x^* \neq 0$. □

Remark 19.4. The following alternative proof for the Problem 11.6 posed by Stephen Simons in [120], which we received anonymously, is worth mentioning.

Assuming that the pair f, g is totally Fenchel unstable, as noticed, we have $\operatorname{dom} f \cap \operatorname{dom} g \neq \emptyset$. As follows from the proof of Proposition 19.2, in case $\operatorname{ri}(\operatorname{dom} f) \cap \operatorname{ri}(\operatorname{dom} g) \neq \emptyset$, f and g satisfy stable Fenchel duality and the conclusion follows. Assume now that $\operatorname{ri}(\operatorname{dom} f) \cap \operatorname{ri}(\operatorname{dom} g) = \emptyset$ and consider an element z in $\operatorname{dom} f \cap \operatorname{dom} g = \operatorname{dom}(f+g)$. By [113, Theorem 11.3] there exists $z^* \in \mathbb{R}^n \setminus \{0\}$ such that

$$\langle z^*, x\rangle \leq \langle z^*, y\rangle \ \forall x \in \operatorname{dom} f \ \forall y \in \operatorname{dom} g,$$

which yields

$$\langle z^*, x\rangle \leq \langle z^*, y\rangle = \langle z^*, z\rangle \ \forall x \in \operatorname{dom} f \ \forall y \in \operatorname{dom} f \cap \operatorname{dom} g.$$

Thus

$$\begin{aligned} f^*(z^* - y^*) &= \sup_{x \in \operatorname{dom} f} \{\langle z^* - y^*, x\rangle - f(x)\} \\ &\leq \langle z^*, z\rangle + \sup_{x \in \operatorname{dom} f} \{\langle -y^*, x\rangle - f(x)\} = \langle z^*, z\rangle + f^*(-y^*) \end{aligned}$$

and so, using (18.2),

$$\begin{aligned} (f+g)^*(z^*) &= \sup_{y \in \operatorname{dom}(f+g)} \{\langle z^*, y\rangle - (f+g)(y)\} \\ &= \langle z^*, z\rangle + \sup_{y \in \operatorname{dom}(f+g)} \{-(f+g)(y)\} = \langle z^*, z\rangle + (f+g)^*(0) \\ &= \langle z^*, z\rangle + f^*(-y^*) + g^*(y^*) \geq f^*(z^* - y^*) + g^*(y^*). \end{aligned}$$

Since $(z^* - y^*) + y^* = z^* \neq 0$, this contradicts the total Fenchel unstability of the pair f, g.

20 Quasi Interior and Quasi-relative Interior

As seen in the previous chapters, for closing the duality gap between the optimal objective value of a convex optimization problem and its Fenchel dual, one needs the fulfillment of a regularity condition. We introduced in Section 2 several generalized interior point conditions for the formulation of which, along the classical interior, the algebraic interior and the strong quasi-relative interior have been used. Unfortunately, for infinite dimensional convex optimization problems, also in practice, it can happen that these duality results cannot be applied because, for instance, the interior of the set involved in the regularity condition is empty. This is the case, for example, when we deal with the positive cones

$$\ell^p_+ = \{x = (x_n)_{n \in \mathbb{N}} \in \ell^p : x_n \geq 0 \ \forall n \in \mathbb{N}\}$$

and

$$L_+^p(T,\mu) = \{u \in L^p(T,\mu) : u(t) \geq 0 \text{ a.e. in } T\}$$

of the spaces l^p and $L^p(T,\mu)$, respectively, where (T,μ) is a σ-finite measure space and $p \in [1,\infty)$. Moreover, also the strong quasi-relative interior of this set, which is the weakest generalized interior notion considered in this context, is empty. For this reason, for a convex set Borwein and Lewis introduced in [11] the notion of quasi-relative interior, which generalizes all the above mentioned interior notions. They proved that the quasi-relative interiors of ℓ_+^p and $L_+^p(T,\mu)$ are nonempty. The same happens also with the quasi interiors of these cones. In the next section, we give for the primal–dual pair (P^{id})–(D^{id}) sufficient conditions, formulated via the quasi interior and the quasi-relative interior, which guarantee strong duality.

To this end, we introduce in this section some preliminary notions and prove several helpful results. Consider X a separated locally convex space and let X^* be its topological dual space. One can easily prove that for a nonempty convex set $U \subseteq X$ it holds

$$\operatorname{coneco}(U \cup \{0\}) = \operatorname{cone}(U). \tag{20.1}$$

The *quasi-relative interior* of U is the set (see [11])

$$\operatorname{qri}(U) = \{u \in U : \operatorname{cl}(\operatorname{cone}(U-u)) \text{ is a linear subspace of } X\}.$$

Next we give an useful characterization of the quasi-relative interior of a convex set, the proof of which can be found in [11].

Proposition 20.1. *Let U be a convex subset of X and $u \in U$. Then $u \in \operatorname{qri}(U)$ if and only if $N_U(u)$ is a linear subspace of X^*.*

In the following, we consider another interior notion for a convex set which is close to the one of quasi-relative interior. For U being a convex subset of X the *quasi interior* of U is the set

$$\operatorname{qi}(U) = \{u \in U : \operatorname{cl}(\operatorname{cone}(U-u)) = X\}.$$

The following characterization of the quasi interior of a convex set was given in [60], where X was considered a reflexive Banach space. As follows from the result below, this property is true even in a separated locally convex vector space.

Proposition 20.2. *Let U be a convex subset of X and $u \in U$. Then $u \in \operatorname{qi}(U)$ if and only if $N_U(u) = \{0\}$.*

Proof. Assume first that $u \in \operatorname{qi}(U)$ and take an arbitrary element $x^* \in N_U(u)$. One can easily see that $\langle x^*, x\rangle \leq 0$ for all $x \in \operatorname{cl}(\operatorname{cone}(U-u))$. Thus $\langle x^*, x\rangle \leq 0$ for all $x \in X$, which is nothing else than $x^* = 0$.

In order to prove the opposite implication, we consider an arbitrary $x \in X$ and prove that $x \in \operatorname{cl}(\operatorname{cone}(U-u))$. Assuming the contrary, by a separation theorem (see for instance [127, Theorem 1.1.5]), one has that there exist $x^* \in X^* \setminus \{0\}$ and $\alpha \in \mathbb{R}$ such that

$$\langle x^*, z\rangle < \alpha < \langle x^*, x\rangle \ \forall z \in \mathrm{cl}(\mathrm{cone}(U - u)).$$

Let $y \in U$ be fixed. For all $\lambda > 0$ it holds $\langle x^*, y - u\rangle < (1/\lambda)\alpha$ and this implies that $\langle x^*, y - u\rangle \leq 0$. As this inequality is true for every arbitrary $y \in U$, we obtain that $x^* \in N_U(u)$. But this leads to a contradiction and in this way the conclusion follows. □

From the definitions of the generalized interior notions introduced above one has that $\mathrm{core}(U) \subseteq \mathrm{sqri}(U) \subseteq \mathrm{qri}(U)$ and $\mathrm{core}(U) \subseteq \mathrm{qi}(U) \subseteq \mathrm{qri}(U)$, while $\mathrm{qri}(\{x\}) = \{x\}$ for all $x \in X$. Moreover, if $\mathrm{qi}(U) \neq \emptyset$, then $\mathrm{qi}(U) = \mathrm{qri}(U)$. This property is given in [93] in the case of a real normed space, but it holds also in a separated locally convex vector space. If X is finite dimensional, then $\mathrm{sqri}(U) = \mathrm{icr}(U) = \mathrm{qri}(U) = \mathrm{ri}(U)$ (cf. [11, 76]) and $\mathrm{int}(U) = \mathrm{core}(U) = \mathrm{qi}(U)$ (cf. [93, 116]). Some useful properties of the quasi-relative interior are listed below. For the proof of (i)–(viii), we refer to [9, 11].

Proposition 20.3. *Let us consider U and V two convex subsets of X, $x \in X$ and $\alpha \in \mathbb{R} \setminus \{0\}$. Then:*

(i) $\mathrm{qri}(U) + \mathrm{qri}(V) \subseteq \mathrm{qri}(U + V)$;
(ii) $\mathrm{qri}(U \times V) = \mathrm{qri}(U) \times \mathrm{qri}(V)$;
(iii) $\mathrm{qri}(U - x) = \mathrm{qri}(U) - x$;
(iv) $\mathrm{qri}(\alpha U) = \alpha\, \mathrm{qri}(U)$;
(v) $t\, \mathrm{qri}(U) + (1 - t)U \subseteq \mathrm{qri}(U)$ *for all* $t \in (0, 1]$, *hence* $\mathrm{qri}(U)$ *is a convex set;*
(vi) *if U is an affine set then* $\mathrm{qri}(U) = U$;
(vii) $\mathrm{qri}(\mathrm{qri}(U)) = \mathrm{qri}(U)$.

If $\mathrm{qri}(U) \neq \emptyset$ *then:*

(viii) $\mathrm{cl}(\mathrm{qri}(U)) = \mathrm{cl}(U)$;
(ix) $\mathrm{cl}(\mathrm{cone}(\mathrm{qri}(U))) = \mathrm{cl}(\mathrm{cone}(U))$.

Proof. (ix) The inclusion $\mathrm{cl}(\mathrm{cone}(\mathrm{qri}(U))) \subseteq \mathrm{cl}(\mathrm{cone}(U))$ is obvious. We prove that $\mathrm{cone}(U) \subseteq \mathrm{cl}(\mathrm{cone}(\mathrm{qri}(U)))$. Consider $x \in \mathrm{cone}(U)$ arbitrary. There exist $\lambda \geq 0$ and $u \in U$ such that $x = \lambda u$. Take $v \in \mathrm{qri}(U)$. Applying the property (v) we get $tv + (1 - t)u \in \mathrm{qri}(U)$ for all $t \in (0, 1]$, so $\lambda tv + (1 - t)x = \lambda(tv + (1 - t)u) \in \mathrm{cone}(\mathrm{qri}(U))$ for all $t \in (0, 1]$. Passing to the limit as $t \searrow 0$ we obtain $x \in \mathrm{cl}(\mathrm{cone}(\mathrm{qri}(U)))$ and hence the desired conclusion follows. □

The next lemma plays an important role for the investigations we make in this chapter.

Lemma 20.4. *Let U and V be nonempty convex subsets of X such that* $\mathrm{qri}(U) \cap V \neq \emptyset$. *If* $0 \in \mathrm{qi}(U - U)$, *then* $0 \in \mathrm{qi}(U - V)$.

Proof. Take $x \in \mathrm{qri}(U) \cap V$ and let $x^* \in N_{U-V}(0)$ be arbitrary. We get $\langle x^*, u - v\rangle \leq 0$ for all $u \in U$ and $v \in V$. This implies

$$\langle x^*, u - x\rangle \leq 0 \ \forall u \in U, \tag{20.2}$$

that is $x^* \in N_U(x)$. As $x \in \text{qri}(U)$, $N_U(x)$ is a linear subspace of X^*, hence $-x^* \in N_U(x)$, which is nothing else than

$$\langle x^*, x - u \rangle \leq 0 \ \forall u \in U. \tag{20.3}$$

The relations (20.2) and (20.3) give us $\langle x^*, u' - u'' \rangle \leq 0$ for all $u', u'' \in U$, so $x^* \in N_{U-U}(0)$. Since $0 \in \text{qi}(U - U)$ we have $N_{U-U}(0) = \{0\}$ (cf. Proposition 20.2) and we get $x^* = 0$. As x^* was arbitrary chosen, we obtain $N_{U-V}(0) = \{0\}$ and, by using again Proposition 20.2, the conclusion follows. □

The following result is a direct consequence of the lemma above.

Lemma 20.5. *Let U be a nonempty convex subset of X and $u \in X$. Then $u \in \text{qri}(U)$ and $0 \in \text{qi}(U - U)$ if and only if $u \in \text{qi}(U)$.*

Proof. The necessity follows from Lemma 20.4 by taking $V = \{u\}$. For getting the sufficiency, it is enough to notice that in case $u \in \text{qi}(U)$, then automatically $u \in \text{qri}(U)$. Moreover, we have $U - u \subseteq U - U$ and since $\text{cl}(\text{cone}(U - u)) = X$, it yields $\text{cl}(\text{cone}(U - U)) = X$ and therefore $0 \in \text{qi}(U - U)$. □

Next, we give useful separation theorems in terms of the notion of quasi-relative interior.

Theorem 20.6. *Let U be a convex subset of X and $u \in U$. If $u \notin \text{qri}(U)$, then there exists $x^* \in X^*, x^* \neq 0$, such that*

$$\langle x^*, x \rangle \leq \langle x^*, u \rangle \ \forall x \in U.$$

Viceversa, if there exists $x^* \in X^*$, $x^* \neq 0$, such that $\langle x^*, x \rangle \leq \langle x^*, u \rangle$ for all $x \in U$ and $0 \in \text{qi}(U - U)$, then $u \notin \text{qri}(U)$.

Proof. Suppose that $u \notin \text{qri}(U)$. According to Proposition 20.1, $N_U(u)$ is not a linear subspace of X^*, hence there exists $x^* \in N_U(u)$, $x^* \neq 0$. Using the definition of the normal cone, we get that $\langle x^*, x \rangle \leq \langle x^*, u \rangle$ for all $x \in U$.

Conversely, assume that there exists $x^* \in X^*$, $x^* \neq 0$, such that $\langle x^*, x \rangle \leq \langle x^*, u \rangle$ for all $x \in U$ and $0 \in \text{qi}(U - U)$. We obtain

$$\langle x^*, x - u \rangle \leq 0 \ \forall x \in U, \tag{20.4}$$

that is $x^* \in N_U(u)$. If we suppose that $u \in \text{qri}(U)$, then by Lemma 20.5, $u \in \text{qi}(U)$ and further, by Proposition 20.2, it yields that $x^* = 0$. This is a contradiction and so $u \notin \text{qri}(U)$. □

Remark 20.7. In [59, 60] a similar separation theorem in case when X is a real normed space is given. For the second part of the above theorem, the authors require that the following condition must be fulfilled: $\text{cl}(T_U(u) - T_U(u)) = X$, where

$$T_U(u) = \left\{ y \in X : y = \lim_{n \to \infty} \lambda_n (u_n - u), \lambda_n > 0, u_n \in U \ \forall n \in \mathbb{N}, \lim_{n \to \infty} u_n = u \right\}$$

is called the *Bouligand tangent cone* to U at $u \in U$. In general, we have the following inclusion $T_U(u) \subseteq \text{cl}(\text{cone}(U - u))$. If the set U is convex, then $T_U(u) = \text{cl}(\text{cone}(U-u))$ (cf. [81]). Using that $\text{cone}(U)-\text{cone}(U) = \text{cone}(U-U)$, if U is a convex subset of X such that $0 \in U$, the condition $\text{cl}(T_U(u)-T_U(u)) = X$ can be reformulated as $\text{cl}(\text{cone}(U - U)) = X$ or, equivalently, $0 \in \text{qi}(U - U)$. Indeed, we have $\text{cl}\,(\text{cl}(\text{cone}(U - u)) - \text{cl}(\text{cone}(U - u))) = X \Leftrightarrow \text{cl}(\text{cone}(U-u)-\text{cone}(U - u)) = X \Leftrightarrow \text{cl}(\text{cone}(U - U)) = X \Leftrightarrow 0 \in \text{qi}(U - U)$. This means that Theorem 20.6 extends to separated locally convex spaces the separation theorem given in [59, 60].

The condition $u \in U$ in Theorem 20.6 is essential (see [60]). However, if u is an arbitrary element of X, it is still possible to give a separation theorem by means of the following result due to Cammaroto and Di Bella (see [54, Theorem 2.1]).

Theorem 20.8 (cf. [54]). *Let U and V be nonempty convex subsets of X with* $\text{qri}(U) \neq \emptyset$, $\text{qri}(V) \neq \emptyset$ *and such that* $\text{cl}(\text{cone}(\text{qri}(U) - \text{qri}(V)))$ *is not a linear subspace of X. Then there exists $x^* \in X^*$, $x^* \neq 0$, such that $\langle x^*, u\rangle \leq \langle x^*, v\rangle$ for all $u \in U$ and $v \in V$.*

The following result is a direct consequence of Theorem 20.8.

Corollary 20.9. *Let U be a convex subset of X and $v \in X$ such that* $\text{qri}(U) \neq \emptyset$ *and* $\text{cl}(\text{cone}(U - v))$ *is not a linear subspace of X. Then there exists $x^* \in X^*, x^* \neq 0$, such that $\langle x^*, u\rangle \leq \langle x^*, v\rangle$ for all $u \in U$.*

Proof. We take in Theorem 20.8 $V := \{u\}$. Then we apply Proposition 20.3(iii), (ix) to obtain the conclusion. □

21 Regularity Conditions via qi and qri

Consider $f, g : X \to \overline{\mathbb{R}}$ two proper and convex functions such that $\text{dom}\, f \cap \text{dom}\, g \neq \emptyset$. The question if whenever for having strong duality for

$$(P^{\text{id}}) \quad \inf_{x \in X} \{f(x) + g(x)\}$$

and its Fenchel dual

$$(D^{\text{id}}) \quad \sup_{y^* \in X^*} \{-f^*(-y^*) - g^*(y^*)\}$$

one can replace in the formulation of the regularity conditions (RC_i^{id}), $i \in \{2, 2', 2''\}$, the generalized interior notions by the quasi-relative interior was addressed by Gowda and Teboulle in [76]. More precisely, whenever X is a Fréchet space and one additionally assumes that f and g are lower semicontinuous, is the condition $0 \in \text{qri}(\text{dom}\, g - \text{dom}\, f)$ sufficient for strong duality for the primal–dual pair

$(P^{\mathrm{id}})-(D^{\mathrm{id}})$? The following example, which can be found in [76], gives us a negative answer and this means that we need additional assumptions in order to guarantee Fenchel duality.

Example 21.1. Take $X = \ell^2$ the Hilbert space consisting of all sequences $x = (x_n)_{n\in\mathbb{N}}$ such that $\sum_{n=1}^{\infty} x_n^2 < \infty$. Consider also the sets

$$C = \{x \in \ell^2 : x_{2n-1} + x_{2n} = 0\ \forall n \in \mathbb{N}\}$$

and

$$S = \{x \in \ell^2 : x_{2n} + x_{2n+1} = 0\ \forall n \in \mathbb{N}\},$$

which are closed linear subspaces of ℓ^2 and fulfill $C \cap S = \{0\}$. Define the functions $f, g : \ell^2 \to \overline{\mathbb{R}}$ by $f = \delta_C(x)$ and $g(x) = x_1 + \delta_S(x)$. Then f and g are proper, convex and lower semicontinuous functions with $\operatorname{dom} f = C$ and $\operatorname{dom} g = S$. As $v(P^{\mathrm{id}}) = 0$ and $v(D^{\mathrm{id}}) = -\infty$ (see [76]), we have a duality gap between the optimal objective values of the primal and dual problem. Moreover, $S - C$ is dense in ℓ^2, thus $\operatorname{cl}(\operatorname{cone}(\operatorname{dom} f - \operatorname{dom} g)) = \operatorname{cl}(C - S) = \ell^2$, which implies $0 \in \operatorname{qi}(\operatorname{dom} f - \operatorname{dom} g)$, hence $0 \in \operatorname{qri}(\operatorname{dom} f - \operatorname{dom} g)$.

In this section we provide sufficient conditions expressed via the quasi interior and quasi-relative interior that guarantee strong duality for (P^{id}) and its Fenchel dual. As in case $v(P^{\mathrm{id}}) = -\infty$ this is automatically fulfilled, we suppose in the following that $v(P^{\mathrm{id}}) \in \mathbb{R}$. First, let us notice that for a function $k : X \to \overline{\mathbb{R}}$ and a real number $\alpha \in \mathbb{R}$, we denote by $k - \alpha : X \to \overline{\mathbb{R}}$ the function $(k-\alpha)(x) = k(x) - \alpha$, while $\widehat{\operatorname{epi}}\, k := \{(x, r) \in X \times \mathbb{R} : (x, -r) \in \operatorname{epi} k\}$ is the *symmetric of* $\operatorname{epi} k$ with respect to the x-axis.

Lemma 21.2. *The following relation is always true*

$$0 \in \operatorname{qri}(\operatorname{dom} f - \operatorname{dom} g) \Rightarrow (0, 1) \in \operatorname{qri}\left(\operatorname{epi} f - \widehat{\operatorname{epi}}(g - v(P^{\mathrm{id}}))\right).$$

Proof. One can see that $\widehat{\operatorname{epi}}(g - v(P^{\mathrm{id}})) = \{(x, r) \in X \times \mathbb{R} : r \le -g(x) + v(P^{\mathrm{id}})\}$. Let us prove first that $(0, 1) \in \operatorname{epi} f - \widehat{\operatorname{epi}}(g - v(P^{\mathrm{id}}))$. Since $\inf_{x\in X}\{f(x) + g(x)\} = v(P^{\mathrm{id}}) < v(P^{\mathrm{id}}) + 1$, there exists $x' \in X$ such that $f(x') + g(x') < v(P^{\mathrm{id}}) + 1$. Then $(0, 1) = \left(x', v(P^{\mathrm{id}}) + 1 - g(x')\right) - \left(x', -g(x') + v(P^{\mathrm{id}})\right) \in \operatorname{epi} f - \widehat{\operatorname{epi}}(g - v(P^{\mathrm{id}}))$.

Now let $(x^*, r^*) \in N_{\operatorname{epi} f - \widehat{\operatorname{epi}}(g - v(P^{\mathrm{id}}))}(0, 1)$. We have

$$\langle x^*, x - x'\rangle + r^*(\mu - \mu' - 1) \le 0\ \forall (x, \mu) \in \operatorname{epi} f\ \forall (x', \mu') \in \widehat{\operatorname{epi}}(g - v(P^{\mathrm{id}})). \tag{21.1}$$

For $(x, \mu) := (x_0, f(x_0))$ and $(x', \mu') := (x_0, -g(x_0) + v(P^{\mathrm{id}}) - 2)$ in (21.1), where $x_0 \in \operatorname{dom} f \cap \operatorname{dom} g$ is fixed, we get $r^*(f(x_0) + g(x_0) - v(P^{\mathrm{id}}) + 1) \le 0$, hence $r^* \le 0$. As $\inf_{x\in X}\{f(x) + g(x)\} = v(P^{\mathrm{id}}) < v(P^{\mathrm{id}}) + 1/2$, there exists

$x_1 \in X$ such that $f(x_1) + g(x_1) < v(P^{\mathrm{id}}) + 1/2$. By taking now $(x, \mu) := (x_1, f(x_1))$ and $(x', \mu') := (x_1, -g(x_1) + v(P^{\mathrm{id}}) - 1/2)$ in (21.1) we obtain $r^*(f(x_1)+g(x_1)-v(P^{\mathrm{id}})-1/2) \leq 0$ and so $r^* \geq 0$. Thus $r^* = 0$ and (21.1) gives $\langle x^*, x - x'\rangle \leq 0$ for all $x \in \operatorname{dom} f$ and $x' \in \operatorname{dom} g$. Hence $x^* \in N_{\operatorname{dom} f - \operatorname{dom} g}(0)$. Since $N_{\operatorname{dom} f - \operatorname{dom} g}(0)$ is a linear subspace of X^* (cf. Proposition 20.2), we have $\langle -x^*, x - x'\rangle \leq 0$ for all $x \in \operatorname{dom} f$ and $x' \in \operatorname{dom} g$ and so $-(x^*, r^*) = (-x^*, 0) \in N_{\operatorname{epi} f - \widehat{\operatorname{epi}}(g - v(P^{\mathrm{id}}))}(0, 1)$, showing that $N_{\operatorname{epi} f - \widehat{\operatorname{epi}}(g - v(P^{\mathrm{id}}))}(0, 1)$ is a linear subspace of $X^* \times \mathbb{R}$. Hence, by applying again Proposition 20.2, we get $(0, 1) \in \operatorname{qri}\left(\operatorname{epi} f - \widehat{\operatorname{epi}}(g - v(P^{\mathrm{id}}))\right)$. □

Proposition 21.3. *Assuming that* $0 \in \operatorname{qi}[(\operatorname{dom} f - \operatorname{dom} g) - (\operatorname{dom} f - \operatorname{dom} g)]$, $N_{\operatorname{co}[(\operatorname{epi} f - \widehat{\operatorname{epi}}(g - v(P^{\mathrm{id}}))) \cup \{(0,0)\}]}(0, 0)$ *is a linear subspace of* $X^* \times \mathbb{R}$ *if and only if* $N_{\operatorname{co}[(\operatorname{epi} f - \widehat{\operatorname{epi}}(g - v(P^{\mathrm{id}}))) \cup \{(0,0)\}]}(0, 0) = \{(0, 0)\}$.

Proof. The sufficiency is trivial. Suppose that $N_{\operatorname{co}[(\operatorname{epi} f - \widehat{\operatorname{epi}}(g - v(P^{\mathrm{id}}))) \cup \{(0,0)\}]}(0, 0)$ is a linear subspace of $X^* \times \mathbb{R}$. Take $(x^*, r^*) \in N_{\operatorname{co}[(\operatorname{epi} f - \widehat{\operatorname{epi}}(g - v(P^{\mathrm{id}}))) \cup \{(0,0)\}]}(0, 0)$. Then

$$\langle x^*, x - x'\rangle + r^*(\mu - \mu') \leq 0 \ \forall (x, \mu) \in \operatorname{epi} f \ \forall (x', \mu') \in \widehat{\operatorname{epi}}(g - v(P_F)). \quad (21.2)$$

Let $x_0 \in \operatorname{dom} f \cap \operatorname{dom} g$ be fixed. Taking $(x, \mu) := (x_0, f(x_0)) \in \operatorname{epi} f$ and $(x', \mu') := (x_0, -g(x_0) + v(P^{\mathrm{id}}) - 1/2) \in \widehat{\operatorname{epi}}(g - v(P^{\mathrm{id}}))$ in the previous inequality we get $r^*(f(x_0) + g(x_0) - v(P^{\mathrm{id}}) + 1/2) \leq 0$, implying $r^* \leq 0$. As $N_{\operatorname{co}[(\operatorname{epi} f - \widehat{\operatorname{epi}}(g - v(P^{\mathrm{id}}))) \cup \{(0,0)\}]}(0, 0)$ is a linear subspace of $X^* \times \mathbb{R}$, the same argument also applies for $(-x^*, -r^*)$, implying $-r^* \leq 0$. In this way we get $r^* = 0$. The inequality (21.2) and $(-x^*, 0) \in N_{\operatorname{co}[(\operatorname{epi} f - \widehat{\operatorname{epi}}(g - v(P^{\mathrm{id}}))) \cup \{(0,0)\}]}(0, 0)$ imply

$$\langle x^*, x - x'\rangle = 0 \ \forall (x, \mu) \in \operatorname{epi} f \ \forall (x', \mu') \in \widehat{\operatorname{epi}}(g - v(P^{\mathrm{id}})),$$

which is nothing else than $\langle x^*, x - x'\rangle = 0$ for all $x \in \operatorname{dom} f$ and $x' \in \operatorname{dom} g$, thus $\langle x^*, x\rangle = 0$ for all $x \in \operatorname{dom} f - \operatorname{dom} g$. Since x^* is linear and continuous, one has that $\langle x^*, x\rangle = 0$ for all $x \in \operatorname{cl}(\operatorname{cone}((\operatorname{dom} f - \operatorname{dom} g) - (\operatorname{dom} f - \operatorname{dom} g))) = X$, hence $x^* = 0$ and the conclusion follows. □

Remark 21.4. (a) By (20.1) one can see that $\operatorname{cl}\left[\operatorname{cone}\left(\operatorname{epi} f - \widehat{\operatorname{epi}}(g - v(P^{\mathrm{id}}))\right)\right]$ coincides with $\operatorname{cl}\left[\operatorname{coneco}\left((\operatorname{epi} f - \widehat{\operatorname{epi}}(g - v(P^{\mathrm{id}}))) \cup \{(0,0)\}\right)\right]$. Hence one has the following sequence of equivalences: $N_{\operatorname{co}[(\operatorname{epi} f - \widehat{\operatorname{epi}}(g - v(P^{\mathrm{id}}))) \cup \{(0,0)\}]}(0, 0)$ is a linear subspace of $X^* \times \mathbb{R} \Leftrightarrow (0, 0) \in \operatorname{qri}\left[\operatorname{co}\left((\operatorname{epi} f - \widehat{\operatorname{epi}}(g - v(P^{\mathrm{id}}))) \cup \{(0,0)\}\right)\right]$ $\Leftrightarrow \operatorname{cl}\left[\operatorname{coneco}\left((\operatorname{epi} f - \widehat{\operatorname{epi}}(g - v(P^{\mathrm{id}}))) \cup \{(0,0)\}\right)\right]$ is a linear subspace of $X \times \mathbb{R} \Leftrightarrow \operatorname{cl}\left[\operatorname{cone}\left(\operatorname{epi} f - \widehat{\operatorname{epi}}(g - v(P^{\mathrm{id}}))\right)\right]$ is a linear subspace of $X \times \mathbb{R}$. The relation $N_{\operatorname{co}[(\operatorname{epi} f - \widehat{\operatorname{epi}}(g - v(P^{\mathrm{id}}))) \cup \{(0,0)\}]}(0, 0) = \{(0, 0)\}$ is equivalent to (cf. Proposition 20.2) $(0, 0) \in \operatorname{qi}\left[\operatorname{co}\left((\operatorname{epi} f - \widehat{\operatorname{epi}}(g - v(P^{\mathrm{id}}))) \cup \{(0,0)\}\right)\right]$, thus in case

$0 \in \operatorname{qi}[(\operatorname{dom} f - \operatorname{dom} g) - (\operatorname{dom} f - \operatorname{dom} g)]$, the conclusion of the previous proposition can be reformulated as follows:

$$\operatorname{cl}\left[\operatorname{cone}(\operatorname{epi} f - \widehat{\operatorname{epi}}(g - v(P^{\mathrm{id}})))\right] \text{ is a linear subspace of } X \times \mathbb{R}$$

$$\Leftrightarrow (0,0) \in \operatorname{qi}\left[\operatorname{co}\left((\operatorname{epi} f - \widehat{\operatorname{epi}}(g - v(P^{\mathrm{id}}))) \cup \{(0,0)\}\right)\right]$$

or, equivalently,

$$(0,0) \in \operatorname{qri}\left[\operatorname{co}\left((\operatorname{epi} f - \widehat{\operatorname{epi}}(g - v(P^{\mathrm{id}}))) \cup \{(0,0)\}\right)\right]$$

$$\Leftrightarrow (0,0) \in \operatorname{qi}\left[\operatorname{co}\left((\operatorname{epi} f - \widehat{\operatorname{epi}}(g - v(P^{\mathrm{id}}))) \cup \{(0,0)\}\right)\right].$$

(b) One can prove that the primal problem (P^{id}) has an optimal solution if and only if $(0,0) \in \operatorname{epi} f - \widehat{\operatorname{epi}}(g - v(P^{\mathrm{id}}))$. This means that if we suppose that the primal problem has an optimal solution and $0 \in \operatorname{qi}[(\operatorname{dom} f - \operatorname{dom} g) - (\operatorname{dom} f - \operatorname{dom} g)]$, then the conclusion of the previous proposition can be rewritten as follows: $N_{(\operatorname{epi} f - \widehat{\operatorname{epi}}(g - v(P^{\mathrm{id}})))}(0,0)$ is a linear subspace of $X^* \times \mathbb{R}$ if and only if $N_{(\operatorname{epi} f - \widehat{\operatorname{epi}}(g - v(P^{\mathrm{id}})))}(0,0) = \{(0,0)\}$ or, equivalently,

$$(0,0) \in \operatorname{qri}\left[\operatorname{epi} f - \widehat{\operatorname{epi}}(g - v(P^{\mathrm{id}}))\right] \Leftrightarrow (0,0) \in \operatorname{qi}\left[\operatorname{epi} f - \widehat{\operatorname{epi}}(g - v(P^{\mathrm{id}}))\right].$$

We give now a first strong duality result for (P^{id}) and its Fenchel dual (D^{id}). Let us notice that for the functions f and g only convexity but no topological assumptions are supposed (see [20]). The fact that the hypotheses $0 \in \operatorname{qi}[(\operatorname{dom} f - \operatorname{dom} g) - (\operatorname{dom} f - \operatorname{dom} g)]$ and $0 \in \operatorname{qri}(\operatorname{dom} f - \operatorname{dom} g)$ can be replaced by $0 \in \operatorname{qi}(\operatorname{dom} f - \operatorname{dom} g)$ follows by Lemma 20.5.

Theorem 21.5. *Suppose that* $0 \in \operatorname{qi}[(\operatorname{dom} f - \operatorname{dom} g) - (\operatorname{dom} f - \operatorname{dom} g)]$, $0 \in \operatorname{qri}(\operatorname{dom} f - \operatorname{dom} g)$ *(or, equivalently,* $0 \in \operatorname{qi}(\operatorname{dom} f - \operatorname{dom} g)$*) and*

$$(0,0) \notin \operatorname{qri}\left[\operatorname{co}\left((\operatorname{epi} f - \widehat{\operatorname{epi}}(g - v(P^{\mathrm{id}}))) \cup \{(0,0)\}\right)\right].$$

Then $v(P^{\mathrm{id}}) = v(D^{\mathrm{id}})$ *and* (D^{id}) *has an optimal solution.*

Proof. Lemma 21.2 ensures that $(0,1) \in \operatorname{qri}\left(\operatorname{epi} f - \widehat{\operatorname{epi}}(g - v(P^{\mathrm{id}}))\right)$, hence $\operatorname{qri}\left(\operatorname{epi} f - \widehat{\operatorname{epi}}(g - v(P^{\mathrm{id}}))\right) \neq \emptyset$. The condition

$$(0,0) \notin \operatorname{qri}\left[\operatorname{co}\left((\operatorname{epi} f - \widehat{\operatorname{epi}}(g - v(P^{\mathrm{id}}))) \cup \{(0,0)\}\right)\right]$$

together with the fact that $\operatorname{cl}\left[\operatorname{coneco}\left((\operatorname{epi} f - \widehat{\operatorname{epi}}(g - v(P^{\mathrm{id}}))) \cup \{(0,0)\}\right)\right] = \operatorname{cl}\left[\operatorname{cone}\left(\operatorname{epi} f - \widehat{\operatorname{epi}}(g - v(P^{\mathrm{id}}))\right)\right]$ imply that $\operatorname{cl}\left[\operatorname{cone}\left(\operatorname{epi} f - \widehat{\operatorname{epi}}(g - v(P^{\mathrm{id}}))\right)\right]$ is not a linear subspace of $X \times \mathbb{R}$. We apply Corollary 20.9 with $U := \operatorname{epi} f - \widehat{\operatorname{epi}}(g - v(P^{\mathrm{id}}))$ and $u = (0,0)$. Thus there exists $(x^*, \lambda) \in X^* \times \mathbb{R}$, $(x^*, \lambda) \neq (0,0)$, such that

$$\langle x^*, x\rangle + \lambda\mu \geq \langle x^*, x'\rangle + \lambda\mu' \ \forall (x,\mu) \in \widehat{\operatorname{epi}}(g - v(P^{\mathrm{id}})) \ \forall (x', \mu') \in \operatorname{epi} f. \quad (21.3)$$

We claim that $\lambda \leq 0$. Indeed, if $\lambda > 0$, then for $(x,\mu) := (u, -g(u) + v(P^{\mathrm{id}}))$ and $(x', \mu') := (u, f(u) + n), n \in \mathbb{N}$, where $u \in \operatorname{dom} f \cap \operatorname{dom} g$ is fixed, we obtain from (21.3) that

$$\langle x^*, u\rangle + \lambda(-g(u) + v(P^{\mathrm{id}})) \geq \langle x^*, u\rangle + \lambda(f(u) + n) \ \forall n \in \mathbb{N}.$$

Passing to the limit as $n \to +\infty$ we obtain a contradiction. Next we prove that $\lambda < 0$. Suppose that $\lambda = 0$. Then from (21.3) we have $\langle x^*, x\rangle \geq \langle x^*, x'\rangle$ for all $x \in \operatorname{dom} g$ and $x' \in \operatorname{dom} f$, hence $\langle x^*, x\rangle \leq 0$ for all $x \in \operatorname{dom} f - \operatorname{dom} g$. Using the second part of Theorem 20.6, we obtain $0 \notin \operatorname{qri}(\operatorname{dom} f - \operatorname{dom} g)$, which contradicts the hypothesis. Thus we must have $\lambda < 0$ and so we obtain from (21.3) that

$$\left\langle \frac{1}{\lambda}x^*, x\right\rangle + \mu \leq \left\langle \frac{1}{\lambda}x^*, x'\right\rangle + \mu' \ \forall (x,\mu) \in \widehat{\operatorname{epi}}(g - v(P^{\mathrm{id}})) \ \forall (x', \mu') \in \operatorname{epi} f.$$

Let $r \in \mathbb{R}$ be such that

$$\mu' + \langle x_0^*, x'\rangle \geq r \geq \mu + \langle x_0^*, x\rangle \ \forall (x,\mu) \in \widehat{\operatorname{epi}}(g - v(P^{\mathrm{id}})) \ \forall (x', \mu') \in \operatorname{epi} f,$$

where $x_0^* := (1/\lambda)x^*$. The first inequality shows that $f(x) \geq \langle -x_0^*, x\rangle + r$ for all $x \in X$, that is $f^*(-x_0^*) \leq -r$. The second one gives us $-g(x) + v(P^{\mathrm{id}}) + \langle x_0^*, x\rangle \leq r$ for all $x \in X$, hence $g^*(x_0^*) \leq r - v(P^{\mathrm{id}})$ and so we have $-f^*(-x_0^*) - g^*(x_0^*) \geq r + v(P^{\mathrm{id}}) - r = v(P^{\mathrm{id}})$. This implies that $v(D^{\mathrm{id}}) \geq v(P^{\mathrm{id}})$. As the opposite inequality is always true, we get $v(P^{\mathrm{id}}) = v(D^{\mathrm{id}})$ and x_0^* is an optimal solution of the problem (D^{id}). □

The above theorem combined with Remark 21.4(b) gives us the following result.

Corollary 21.6. *Suppose that the primal problem* (P^{id}) *has an optimal solution,* $0 \in \operatorname{qi}[(\operatorname{dom} f - \operatorname{dom} g) - (\operatorname{dom} f - \operatorname{dom} g)]$, $0 \in \operatorname{qri}(\operatorname{dom} f - \operatorname{dom} g)$ *(or, equivalently,* $0 \in \operatorname{qi}(\operatorname{dom} f - \operatorname{dom} g)$*) and*

$$(0,0) \notin \operatorname{qri}\left[\operatorname{epi} f - \widehat{\operatorname{epi}}(g - v(P^{\mathrm{id}}))\right].$$

Then $v(P^{\mathrm{id}}) = v(D^{\mathrm{id}})$ *and* (D^{id}) *has an optimal solution.*

Some stronger versions of Theorem 21.5 and Corollary 21.6, respectively, follow (see [20]).

Theorem 21.7. *Suppose that* $\operatorname{dom} f \cap \operatorname{qri}(\operatorname{dom} g) \neq \emptyset$, $0 \in \operatorname{qi}(\operatorname{dom} g - \operatorname{dom} g)$ *and*

$$(0,0) \notin \operatorname{qri}\Big[\operatorname{co}\Big((\operatorname{epi} f - \widehat{\operatorname{epi}}(g - v(P^{\mathrm{id}}))) \cup \{(0,0)\}\Big)\Big].$$

Then $v(P^{\mathrm{id}}) = v(D^{\mathrm{id}})$ *and* (D^{id}) *has an optimal solution.*

Proof. We apply Lemma 20.4 with $U := \operatorname{dom} g$ and $V := \operatorname{dom} f$. We get $0 \in \operatorname{qi}(\operatorname{dom} g - \operatorname{dom} f)$ or, equivalently, $0 \in \operatorname{qi}(\operatorname{dom} f - \operatorname{dom} g)$. Now Theorem 21.5 leads to the conclusion. □

Corollary 21.8. *Suppose that the primal problem* (P^{id}) *has an optimal solution,* $\operatorname{dom} f \cap \operatorname{qri}(\operatorname{dom} g) \neq \emptyset$, $0 \in \operatorname{qi}(\operatorname{dom} g - \operatorname{dom} g)$ *and*

$$(0,0) \notin \operatorname{qri}\Big[\operatorname{epi} f - \widehat{\operatorname{epi}}(g - v(P^{\mathrm{id}}))\Big].$$

Then $v(P^{\mathrm{id}}) = v(D^{\mathrm{id}})$ *and* (D^{id}) *has an optimal solution.*

Remark 21.9. (a) In the statements above, we introduce two more regularity conditions for Fenchel duality for which one has that the regularity condition given in Theorem 21.7 (Corollary 21.8) implies the one given in Theorem 21.5 (Corollary 21.6).

(b) If one renounces the condition

$$(0,0) \notin \operatorname{qri}\Big[\operatorname{co}\Big((\operatorname{epi} f - \widehat{\operatorname{epi}}(g - v(P^{\mathrm{id}}))) \cup \{(0,0)\}\Big)\Big]$$

or, respectively, $(0,0) \notin \operatorname{qri}\Big[\operatorname{epi} f - \widehat{\operatorname{epi}}(g - v(P^{\mathrm{id}}))\Big]$, in the case when the primal problem has an optimal solution, then the duality results given above may fail. By using again Example 21.1 we show that these conditions are essential. We notice first that for the problem in Example 21.1 the condition $0 \in \operatorname{qi}(\operatorname{dom} f - \operatorname{dom} g)$ is fulfilled. We prove in the following that in the aforementioned example we have $(0,0) \in \operatorname{qri}\Big[\operatorname{epi} f - \widehat{\operatorname{epi}}(g - v(P^{\mathrm{id}}))\Big]$. Note that the scalar product on ℓ^2, $\langle \cdot, \cdot \rangle : \ell^2 \times \ell^2 \to \mathbb{R}$, is given by $\langle x, y \rangle = \sum_{n=1}^{\infty} x_n y_n$ for $x = (x_n)_{n \in \mathbb{N}}, y = (y_n)_{n \in \mathbb{N}} \in \ell^2$. For $k \in \mathbb{N}$ we denote by $e^{(k)}$ the element in ℓ^2 such that $e_n^{(k)} = 1$, if $n = k$ and $e_n^{(k)} = 0$, for all $n \in \mathbb{N} \setminus \{k\}$. We have $\operatorname{epi} f = C \times [0, +\infty)$. Further, $\widehat{\operatorname{epi}}(g - v(P^{\mathrm{id}})) = \{(x,r) \in \ell^2 \times \mathbb{R} : r \leq -g(x)\} = \{(x,r) \in \ell^2 \times \mathbb{R} : x = (x_n)_{n \in \mathbb{N}} \in S, r \leq -x_1\} = \{(x, -x_1 - \varepsilon) \in \ell^2 \times \mathbb{R} : x = (x_n)_{n \in \mathbb{N}} \in S, \varepsilon \geq 0\}$. Then $U := \operatorname{epi} f - \widehat{\operatorname{epi}}(g - v(P^{\mathrm{id}})) = \{(x - x', x_1' + \varepsilon) : x \in C, x' = (x_n')_{n \in \mathbb{N}} \in S, \varepsilon \geq 0\}$. Take $(x^*, r) \in N_U(0,0)$, where $x^* = (x_n^*)_{n \in \mathbb{N}}$. We have

$$\langle x^*, x - x' \rangle + r(x_1' + \varepsilon) \leq 0 \ \forall x \in C \ \forall x' = (x_n')_{n \in \mathbb{N}} \in S \ \forall \varepsilon \geq 0. \tag{21.4}$$

Taking in (21.4) $x' = 0$ and $\varepsilon = 0$ we get $\langle x^*, x\rangle \leq 0$ for all $x \in C$. As C is a linear subspace of X we have

$$\langle x^*, x\rangle = 0 \ \forall x \in C. \tag{21.5}$$

Since $e^{(2k-1)} - e^{(2k)} \in C$ for all $k \in \mathbb{N}$, relation (21.5) implies

$$x^*_{2k-1} - x^*_{2k} = 0 \ \forall k \in \mathbb{N}. \tag{21.6}$$

From (21.4) and (21.5) we obtain

$$\langle -x^*, x'\rangle + r(x'_1 + \varepsilon) \leq 0 \ \forall x' = (x'_n)_{n\in\mathbb{N}} \in S \ \forall \varepsilon \geq 0. \tag{21.7}$$

By taking $\varepsilon = 0$ and $x' := me^{(1)} \in S$ in (21.7), where $m \in \mathbb{Z}$ is arbitrary, we get $m(-x^*_1 + r) \leq 0$ for all $m \in \mathbb{Z}$, thus $r = x^*_1$. For $\varepsilon = 0$ in (21.7) we obtain $-\sum_{n=1}^{\infty} x^*_n x'_n + r x'_1 \leq 0$ for all $x' \in S$. By taking into account that $r = x^*_1$ we get $-\sum_{n=2}^{\infty} x^*_n x'_n \leq 0$ for all $x' \in S$. As S is a linear subspace of X it follows that $\sum_{n=2}^{\infty} x^*_n x'_n = 0$ for all $x' \in S$, but, since $e^{(2k)} - e^{(2k+1)} \in S$ for all $k \in \mathbb{N}$, the above relation shows that

$$x^*_{2k} - x^*_{2k+1} = 0 \ \forall k \in \mathbb{N}. \tag{21.8}$$

Combining (21.6) with (21.8) we get $x^* = 0$ (since $x^* \in \ell^2$). Because $r = x^*_1$ we also have $r = 0$. Thus $N_U(0,0) = \{(0,0)\}$ and Proposition 20.2 gives us the desired conclusion.

(c) Since in the hypotheses of the strong duality results given above the relation $0 \in \operatorname{qi}[(\operatorname{dom} f - \operatorname{dom} g) - (\operatorname{dom} f - \operatorname{dom} g)]$ is fulfilled, in view of Remark 21.4, the condition $(0,0) \notin \operatorname{qri}\left[\operatorname{co}\left((\operatorname{epi} f - \widehat{\operatorname{epi}}(g - v(P^{\mathrm{id}}))) \cup \{(0,0)\}\right)\right]$ (respectively, $(0,0) \notin \operatorname{qri}\left[\operatorname{epi} f - \widehat{\operatorname{epi}}(g - v(P^{\mathrm{id}}))\right]$) can be replaced by the hypothesis $(0,0) \notin \operatorname{qi}\left[\operatorname{co}\left((\operatorname{epi} f - \widehat{\operatorname{epi}}(g - v(P^{\mathrm{id}}))) \cup \{(0,0)\}\right)\right]$ (respectively, $(0,0) \notin \operatorname{qi}\left[\operatorname{epi} f - \widehat{\operatorname{epi}}(g - v(P^{\mathrm{id}}))\right]$).

(d) We have the following relation

$$(0,0) \in \operatorname{qi}\left[\operatorname{co}\left((\operatorname{epi} f - \widehat{\operatorname{epi}}(g - v(P^{\mathrm{id}}))) \cup \{(0,0)\}\right)\right] \Rightarrow 0 \in \operatorname{qi}(\operatorname{dom} f - \operatorname{dom} g).$$

Indeed, one has $(0,0) \in \operatorname{qi}\left[\operatorname{co}\left((\operatorname{epi} f - \widehat{\operatorname{epi}}(g - v(P^{\mathrm{id}}))) \cup \{(0,0)\}\right)\right]$ if and only if $\operatorname{cl}\left[\operatorname{coneco}\left((\operatorname{epi} f - \widehat{\operatorname{epi}}(g - v(P^{\mathrm{id}}))) \cup \{(0,0)\}\right)\right] = X \times \mathbb{R}$ and, consequently, $\operatorname{cl}\left[\operatorname{cone}\left(\operatorname{epi} f - \widehat{\operatorname{epi}}(g - v(P^{\mathrm{id}}))\right)\right] = X \times \mathbb{R}$. Since

$$\operatorname{cl}\left[\operatorname{cone}\left(\operatorname{epi} f - \widehat{\operatorname{epi}}(g - v(P^{\mathrm{id}}))\right)\right] \subseteq \operatorname{cl}\operatorname{cone}(\operatorname{dom} f - \operatorname{dom} g) \times \mathbb{R},$$

this implies $\text{cl}\,[\text{cone}(\text{dom}\, f - \text{dom}\, g)] = X$, that is $0 \in \text{qi}(\text{dom}\, f - \text{dom}\, g)$. Hence

$$0 \notin \text{qi}(\text{dom}\, f - \text{dom}\, g) \Rightarrow (0,0) \notin \text{qi}\left[\text{co}\left((\text{epi}\, f - \widehat{\text{epi}}(g - v(P^{\text{id}}))) \cup \{(0,0)\}\right)\right].$$

Nevertheless, in the regularity conditions given above one cannot substitute the condition $(0,0) \notin \text{qri}\left[\text{co}\left((\text{epi}\, f - \widehat{\text{epi}}(g - v(P^{\text{id}}))) \cup \{(0,0)\}\right)\right]$ with the "nice-looking" one $0 \notin \text{qi}(\text{dom}\, f - \text{dom}\, g)$, since in all strong duality theorems the hypotheses we consider guarantee $0 \in \text{qi}(\text{dom}\, f - \text{dom}\, g)$.

The following example underlines the applicability of the strong duality theorems introduced in this section.

Example 21.10. Consider again $X = \ell^2$ equipped with the norm $\|\cdot\| : \ell^2 \to \mathbb{R}$, $\|x\|^2 = \sum_{n=1}^{\infty} x_n^2$, $x = (x_n)_{n\in\mathbb{N}} \in \ell^2$. We define the functions $f, g : \ell^2 \to \overline{\mathbb{R}}$ by $f(x) = \|x\| + \delta_{x_0 - \ell^2_+}(x)$ and $g(x) = \langle c, x\rangle + \delta_{\ell^2_+}(x)$, where $\ell^2_+ = \{(x_n)_{n\in\mathbb{N}} \in \ell^2 : x_n \geq 0\ \forall n \in \mathbb{N}\}$ is the positive cone, $x_0 = (1/n)_{n\in\mathbb{N}}$ and $c = (1/2^n)_{n\in\mathbb{N}}$. Note that $v(P^{\text{id}}) = \inf_{x\in\ell^2}\{f(x) + g(x)\} = 0$ and the infimum is attained at $x = 0$. We have $\text{dom}\, f = x_0 - \ell^2_+ = \{(x_n)_{n\in\mathbb{N}} \in \ell^2 : x_n \leq 1/n\ \forall n \in \mathbb{N}\}$ and $\text{dom}\, g = \ell^2_+$. Since $\text{qri}(\ell^2_+) = \{(x_n)_{n\in\mathbb{N}} \in \ell^2 : x_n > 0\ \forall n \in \mathbb{N}\}$ (cf. [11]), we get $\text{dom}\, f \cap \text{qri}(\text{dom}\, g) = \{(x_n)_{n\in\mathbb{N}} \in \ell^2 : 0 < x_n \leq 1/n\ \forall n \in \mathbb{N}\} \neq \emptyset$. Also, $\text{cl}(\text{cone}(\text{dom}\, g - \text{dom}\, g)) = \ell^2$, so $0 \in \text{qi}(\text{dom}\, g - \text{dom}\, g)$. Further, $\text{epi}\, f = \{(x,r) \in \ell^2 \times \mathbb{R} : x \in x_0 - \ell^2_+, \|x\| \leq r\} = \{(x, \|x\| + \varepsilon) \in \ell^2 \times \mathbb{R} : x \in x_0 - \ell^2_+, \varepsilon \geq 0\}$ and $\widehat{\text{epi}}(g - v(P^{\text{id}})) = \{(x,r) \in \ell^2 \times \mathbb{R} : r \leq -g(x)\} = \{(x,r) \in \ell^2 \times \mathbb{R} : r \leq -\langle c, x\rangle, x \in \ell^2_+\} = \{(x, -\langle c, x\rangle - \varepsilon) : x \in \ell^2_+, \varepsilon \geq 0\}$. We get $\text{epi}\, f - \widehat{\text{epi}}(g - v(P^{\text{id}})) = \{(x - x', \|x\| + \varepsilon + \langle c, x'\rangle + \varepsilon') : x \in x_0 - \ell^2_+, x' \in \ell^2_+, \varepsilon, \varepsilon' \geq 0\} = \{(x - x', \|x\| + \langle c, x'\rangle + \varepsilon) : x \in x_0 - \ell^2_+, x' \in \ell^2_+, \varepsilon \geq 0\}$.

In the following we prove that $(0,0) \notin \text{qri}\left[\text{epi}\, f - \widehat{\text{epi}}(g - v(P^{\text{id}}))\right]$. By assuming the contrary we would have that $\text{cl}\left[\text{cone}\left(\text{epi}\, f - \widehat{\text{epi}}(g - v(P^{\text{id}}))\right)\right]$ is a linear subspace. Since $(0,1) \in \text{cl}\left[\text{cone}\left(\text{epi}\, f - \widehat{\text{epi}}(g - v(P^{\text{id}}))\right)\right]$ (take $x = x' = 0$ and $\varepsilon = 1$), we must have that also $(0,-1)$ belongs to this set. On the other hand, one can see that for all (x,r) in $\text{cl}\left[\text{cone}\left(\text{epi}\, f - \widehat{\text{epi}}(g - v(P^{\text{id}}))\right)\right]$ it holds $r \geq 0$. This leads to the sought contradiction.

Hence the conditions of Corollary 21.8 are fulfilled, thus strong duality holds. Let us notice that the regularity conditions given in Corollary 21.6 are fulfilled, too (see Remark 21.9(a)). On the other hand, ℓ^2 is a Fréchet space (being a Hilbert space), the functions f and g are lower semicontinuous and, as $\text{sqri}(\text{dom}\, f - \text{dom}\, g) = \text{sqri}(x_0 - \ell^2_+) = \emptyset$, none of the regularity conditions (RC_i^{id}), $i \in \{2, 2', 2''\}$, can be applied for this optimization problem. Moreover, the regularity condition (RC_1^{id}) is also not valid. As for all $x^* \in \ell^2$ it holds $g^*(x^*) = \delta_{c - \ell^2_+}(x^*)$ and

$$f^*(-x^*) = \inf_{x_1^* + x_2^* = -x^*}\left\{\|\cdot\|^*(x_1^*) + \delta^*_{x_0 - \ell^2_+}(x_2^*)\right\} = \inf_{\substack{x_1^* + x_2^* = -x^* \\ \|x_1^*\| \leq 1, x_2^* \in \ell^2_+}} \langle x_2^*, x_0\rangle,$$

the optimal objective value of the Fenchel dual problem is

$$v(D_F) = \sup_{\substack{x_2^* \in \ell_+^2 - c - x_1^* \\ \|x_1^*\| \leq 1, x_2^* \in \ell_+^2}} \langle -x_2^*, x_0 \rangle = \sup_{x_2^* \in \ell_+^2} \langle -x_2^*, x_0 \rangle = 0$$

and $x_2^* = 0$ is the optimal solution of the dual.

Remark 21.11. Considering Y another separated locally convex space with the topological dual space Y^*, $A : X \to Y$ a linear continuous operator and $f : X \to \overline{\mathbb{R}}$ and $g : Y \to \overline{\mathbb{R}}$ proper and convex functions such that $A(\operatorname{dom} f) \cap \operatorname{dom} g \neq \emptyset$, one can derive from the strong duality results given above corresponding statements for the primal–dual pair (P^A)–(D^A). To this end, one has to define $F, G : X \times Y \to \overline{\mathbb{R}}$, $F(x, y) = f(x) + \delta_{\{u \in X : Au = y\}}(x)$ and $G(x, y) = g(y)$ and to notice that

$$\inf_{(x,y) \in X \times Y} \{F(x, y) + G(x, y)\} = \inf_{x \in X} \{f(x) + (g \circ A)(x)\} = v(P^A)$$

and

$$\begin{aligned} \sup_{(x^*,y^*) \in X^* \times Y^*} \{-F^*(-x^*, -y^*) - G^*(x^*, y^*)\} &= \sup_{y^* \in Y^*} \{-f^*(-A^* y^*) - g^*(y^*)\} \\ &= v(D^A). \end{aligned}$$

Therefore, one can formulate regularity conditions that ensure strong duality for (P^A) and (D^A) by using as starting point the Theorems 21.5 and 21.7 (respectively, the Corollaries 21.6 and 21.8) applied to F and G. For more details we refer to [20]. Let us notice that for the primal–dual pair (P^A)–(D^A) Borwein–Lewis gave in [11] sufficient conditions for strong duality via the quasi-relative interior, but under the very restrictive assumption that Y is finite dimensional.

22 Lagrange Duality via Fenchel Duality

We use an approach due to Magnanti (cf. [95]) and derive in this section regularity conditions expressed via the quasi interior and the quasi-relative interior for the problem with geometric and cone constraints. As starting point we use the achievements from Section 21. The setting we consider here for the primal problem is different to the one in the previous sections, where this problem was treated, as we prefer to work under similar hypotheses like in the literature dealing with the same topic (cf [54, 59, 60]).

Let X be a topological vector space and S a nonempty subset of X. Let Z be a separated locally convex space partially ordered by a nonempty convex cone $C \subseteq Z$. Let $f : S \to \mathbb{R}$ and $g : S \to Z$ be two functions such that the function $(f, g) : S \to \mathbb{R} \times Z$, defined by $(f, g)(x) = (f(x), g(x))$, is *convexlike* with respect to the cone $\mathbb{R}_+ \times C \subseteq \mathbb{R} \times Z$, that is the set $(f, g)(S) + \mathbb{R}_+ \times C$ is convex. This property implies that the sets $f(S) + [0, +\infty)$ and $g(S) + C$ are convex, while the reverse implication does not always hold. Let the feasible set $\mathcal{A} = \{x \in S : g(x) \in -C\}$ be nonempty. For having strong duality for

$$(P^C) \inf_{x \in \mathcal{A}} f(x),$$
$$\mathcal{A} = \{x \in S : g(x) \in -C\}$$

and its Lagrange dual

$$(D^{C_L}) \quad \sup_{z^* \in C^*} \inf_{x \in S} \{f(x) + (z^* g)(x)\},$$

in [54, Theorem 2.2], the following conditions have been stated: $\text{cl}(C - C) = Z$, there exists $x' \in S$ such that $g(x') \in -\text{qri}(C)$, $\text{qri}(g(S) + C) \neq \emptyset$ and $\text{cl}\,[\text{cone}\,(\text{qri}(g(S) + C))]$ is not a linear subspace of Z. Like it results from the following lemma, these assumptions are self-contradictory and this fact motivates the need for providing valuable regularity conditions for the primal–dual pair $(P^C) - (D^{C_L})$ expressed via the quasi interior and quasi-relative interior.

Lemma 22.1. *Suppose that* $\text{cl}(C - C) = Z$ *and there exists* $x' \in S$ *such that* $g(x') \in -\text{qri}(C)$. *Then the following assertions are true:*

(i) $0 \in \text{qi}(g(S) + C)$;
(ii) $\text{cl}\,[\text{cone}\,(\text{qri}(g(S) + C))] = Z$.

Proof. (i) We apply Lemma 20.4 with $U := -C$ and $V := g(S)+C$. The condition $\text{cl}(C - C) = Z$ implies that $0 \in \text{qi}(U - U)$, while the Slater-type condition $g(x') \in -\text{qri}(C)$ ensures that $g(x') \in \text{qri}(U) \cap V$. Hence, by Lemma 20.4, we obtain $0 \in \text{qi}(U - V)$, that is $0 \in \text{qi}(-g(S) - C)$, which is nothing else than $0 \in \text{qi}(g(S) + C)$.

(ii) From (i) it follows that $0 \in \text{qri}(g(S) + C)$. By applying Proposition 20.3(ix), we get $\text{cl}\,[\text{cone}\,(\text{qri}(g(S) + C))] = \text{cl}[\text{cone}(g(S) + C)]$. Using again (i) one has that these sets are nothing else than the whole space Z. □

In the following example we underline the applicability of the result above beyond its importance for the results developed in this section.

Example 22.2. Let $A = [a, b] \subseteq [0, 1]$, $a < b$ and $\bar{v}, \bar{w} \in L^2([0, 1])$ be such that $0 < \bar{w} < \bar{v}$ a.e. in A. Consider $X = Z = L^2([0, 1])$, $C = \{u \in L^2([0, 1]) : u \geq 0 \text{ a.e. in } [0, 1]\}$ and $S = \{u \in L^2([0, 1]) : u|_A = \bar{w}|_A \text{ a.e. in } A\}$. Define the function $g : S \to L^2([0, 1])$ as being $g(u) = u - \bar{v}$. The feasible set $\mathcal{A} = \{u \in S : u \leq \bar{v} \text{ a.e. in } [0, 1]\}$ is nonempty, $\text{cl}(C - C) = L^2([0, 1])$ and $\text{qri}(C) = \{u \in L^2([0, 1]) : u > 0 \text{ a.e. in } [0, 1]\}$. Thus one can easily find an element $x' \in S$ such that $g(x') \in -\text{qri}(C)$ and therefore the hypotheses of Lemma 22.1 are fulfilled, which means that both conclusions (i) and (ii) are true.

Since $\operatorname{qri}(g(S) + C) \neq \emptyset$, by Proposition 20.3(ix) it follows that

$$\operatorname{cl}[\operatorname{cone}(\operatorname{qri}(g(S) + C))] = \operatorname{cl}[\operatorname{cone}(g(S) + C)]$$

$$= \operatorname{cl}[\operatorname{cone}(\{u \in L^2([0,1]) : u|_A \geq \bar{w}|_A - \bar{v}|_A \text{ a.e. in } A\})]$$

and this is, by Lemma 22.1, nothing else than the whole space $L^2([0,1])$.

Next, we show this in an alternative way. Let $z \in L^2([0,1]$ be such that $z|_A := \bar{w}|_A - \bar{v}|_A$. One has that $z|_A < 0$ a.e. in A. Denote by $T := \operatorname{cone}(\{u \in L^2([0,1]) : u|_A \geq z|_A \text{ a.e. in } A\})$ and notice that $T = \cup_{s>0}\{u \in L^2([0,1]) : u|_A \geq sz|_A \text{ a.e. in } A\}$. We prove that $\operatorname{cl}(T) = L^2([0,1])$ and to this end we consider an arbitrary $f \in L^2([0,1])$. For all $n \in \mathbb{N}$ we define $f_n : [0,1] \to \mathbb{R}$ by

$$f_n(x) = \begin{cases} f(x), & x \in ([0,1] \setminus A) \cup \{x \in A : f(x) \geq n\alpha(x)\}, \\ 0, & \text{otherwise.} \end{cases}$$

Obviously, for all $n \in \mathbb{N}$ one has $f_n \in T$. We define $g_n := (f - f_n)^2 \in L^1([0,1])$ for all $n \in N$. For all $x \in [0,1]$ we have that there exists $n_0 \in N$ such that for all $n \geq n_0$ it holds $f_n(x) = f(x)$. Thus, for all $x \in [0,1]$, $g_n(x) \to 0$ $(n \to +\infty)$ (pointwise convergence). On the other hand, for all $n \geq 1$ and $x \in [0,1]$ it holds $|g_n(x)| \leq |f(x)|^2$. One can notice that $|f|^2$ belongs to $L^1([0,1])$ and this allows us to apply Lebesgue's Dominated Convergence Theorem, which guarantees that

$$\lim_{n\to\infty} \|f - f_n\|^2 = \lim_{n\to\infty} \int_0^1 |f(x) - f_n(x)|^2 dx = \lim_{n\to\infty} \int_0^1 |g_n(x)| dx = 0.$$

Consequently, $f \in \operatorname{cl}(T)$ and this means that

$$\operatorname{cl}[\operatorname{cone}(\{u \in L^2([0,1]) : u|_A \geq \bar{w}|_A - \bar{v}|_A \text{ a.e. in } A\})] = L^2([0,1]).$$

We give below regularity conditions written in terms of the quasi interior and quasi-relative interior for Lagrange duality. Since otherwise strong duality is automatically ensured, we suppose in the following that $v(P^C)$ is a real number. Consider the following convex set

$$\mathcal{E}_{v(P^C)} = \{(f(x) + \alpha - v(P^C), g(x) + z) : x \in S, \alpha \geq 0, z \in C\} \subseteq \mathbb{R} \times Z.$$

The set $-\mathcal{E}_{v(P^C)}$ is in analogy to the *conic extension*, a notion used by Franco Giannessi in the theory of *image space analysis* (see [73]). One can easily prove that the primal problem (P^C) has an optimal solution if and only if $(0,0) \in \mathcal{E}_{v(P^C)}$. Consider the functions $F, G : \mathbb{R} \times Z \to \overline{\mathbb{R}}$ defined by

$$F(r,z) = \begin{cases} r, & \text{if } (r,z) \in \mathcal{E}_{v(P^C)} + (v(P^C), 0), \\ +\infty, & \text{otherwise} \end{cases}$$

and $G = \delta_{\mathbb{R}\times(-C)}$. It holds

$$\operatorname{dom} F - \operatorname{dom} G = \mathbb{R} \times (g(S) + C). \tag{22.1}$$

Moreover, as pointed out by Magnanti in [95], we have

$$\inf_{(r,z)\in\mathbb{R}\times Z} \{F(r,z) + G(r,z)\} = \inf_{x\in\mathcal{A}} f(x) = v(P^C) \tag{22.2}$$

and

$$\sup_{(r^*,z^*)\in\mathbb{R}\times Z^*} \{-F^*(-r^*,-z^*) - G^*(r^*,z^*)\} = \sup_{z^*\in C^*} \inf_{x\in S}\{f(x) + (z^*g)(x)\}$$
$$= v(D^{C_L}). \tag{22.3}$$

Next, we derive from the strong duality results given for Fenchel duality corresponding strong duality results for Lagrange duality.

Theorem 22.3. *Suppose that* $0 \in \operatorname{qi}[(g(S)+C)-(g(S)+C)]$, $0 \in \operatorname{qri}(g(S)+C)$ *(or, equivalently,* $0 \in \operatorname{qi}(g(S)+C)$*) and* $(0,0) \notin \operatorname{qri}\left[\operatorname{co}(\mathcal{E}_{v(P^C)} \cup \{(0,0)\})\right]$. *Then* $v(P^C) = v(D^{C_L})$ *and* (D^{C_L}) *has an optimal solution.*

Proof. The hypotheses of the theorem and (22.1) imply that $(0,0) \in \operatorname{qi}[(\operatorname{dom} F - \operatorname{dom} G) - (\operatorname{dom} F - \operatorname{dom} G)]$ and $(0,0) \in \operatorname{qri}(\operatorname{dom} F - \operatorname{dom} G)$. Further $\operatorname{epi} F = \{(r,z,s) \in \mathbb{R}\times Z\times\mathbb{R} : (r,z) \in \mathcal{E}_{v(P^C)} + (v(P^C),0), r \leq s\} = \{(f(x)+\alpha, g(x)+z, s) : x\in S, \alpha\geq 0, z\in C, f(x)+\alpha \leq s\}$ and $\widehat{\operatorname{epi}}(G - v(P^C)) = \mathbb{R}\times(-C)\times(-\infty, v(P^C)]$. Thus $\operatorname{epi} F - \widehat{\operatorname{epi}}(G - v(P^C)) = \{(f(x)+\alpha+a, g(x)+z, f(x)+\alpha+\varepsilon - v(P^C)) : x\in S, \alpha\geq 0, a\in\mathbb{R}, z\in C, \varepsilon\geq 0\} = \mathbb{R}\times\{(g(x)+z, f(x)+\alpha - v(P^C)) : x\in S, \alpha\geq 0, z\in C\}$. This means that

$$(0,0,0) \in \operatorname{qri}\left[\operatorname{co}\left((\operatorname{epi} F - \widehat{\operatorname{epi}}(G - v(P^C))) \cup \{(0,0,0)\}\right)\right]$$

if and only if $(0,0) \notin \operatorname{qri}\left[\operatorname{co}(\mathcal{E}_{v(P^C)} \cup \{(0,0)\})\right]$. Applying Theorem 21.5 for F and G, we obtain

$$\inf_{(r,z)\in\mathbb{R}\times Z} \{F(r,z) + G(r,z)\} = \max_{(r^*,z^*)\in\mathbb{R}\times Z^*} \{-F^*(-r^*,-z^*) - G^*(r^*,z^*)\}. \tag{22.4}$$

By (22.2) and (22.3) the conclusion follows. □

Corollary 22.4. *Suppose that the primal problem* (P^C) *has an optimal solution,* $0 \in \operatorname{qi}[(g(S)+C)-(g(S)+C)]$, $0 \in \operatorname{qri}(g(S)+C)$ *(or, equivalently,* $0 \in \operatorname{qi}(g(S)+C)$*) and* $(0,0) \notin \operatorname{qri}(\mathcal{E}_{v(P^C)})$. *Then* $v(P^C) = v(D^{C_L})$ *and* (D^{C_L}) *has an optimal solution.*

Further, like for Fenchel duality, the following stronger, but more handleable, Lagrange duality results can be stated.

Theorem 22.5. *Suppose that* $\mathrm{cl}(C - C) = Z$ *and there exists* $x' \in S$ *such that* $g(x') \in -\mathrm{qri}(C)$. *If* $(0,0) \notin \mathrm{qri}\left[\mathrm{co}(\mathcal{E}_{v(P^C)} \cup \{(0,0)\})\right]$, *then* $v(P^C) = v(D^{C_L})$ *and* (D^{C_L}) *has an optimal solution.*

Proof. The condition $(0,0) \notin \mathrm{qri}\left[\mathrm{co}(\mathcal{E}_{v(P^C)} \cup \{(0,0)\})\right]$ implies that $(0,0,0) \in \mathrm{qri}\left[\mathrm{co}\left((\mathrm{epi}\, F - \widehat{\mathrm{epi}}(G - v(P^C))) \cup \{(0,0,0)\}\right)\right]$ (cf. the proof of Theorem 22.3). Further, we have $\mathrm{dom}\, F \cap \mathrm{qri}(\mathrm{dom}\, G) = \left(\mathcal{E}_{v(P^C)} + (v(P^C),0)\right) \cap \mathrm{qri}\,(\mathbb{R} \times (-C)) = \left(\mathcal{E}_{v(P^C)} + (v(P^C),0)\right) \cap (\mathbb{R} \times (-\mathrm{qri}(C)))$. From the Slater-type condition we obtain that $(f(x'), g(x')) \in \left(\mathcal{E}_{v(P^C)} + (v(P^C),0)\right) \cap (\mathbb{R} \times (-\mathrm{qri}(C)))$ hence $\mathrm{dom}\, F \cap \mathrm{qri}(\mathrm{dom}\, G) \neq \emptyset$. Moreover, $\mathrm{cl}\,[\mathrm{cone}(\mathrm{dom}\, G - \mathrm{dom}\, G)] = \mathbb{R} \times \mathrm{cl}(C - C) = \mathbb{R} \times Z$, hence $(0,0) \in \mathrm{qi}(\mathrm{dom}\, G - \mathrm{dom}\, G)$. By Theorem 21.7, we obtain (22.4) and by using again (22.2) and (22.3) the conclusion follows. □

Corollary 22.6. *Suppose that the primal problem* (P^C) *has an optimal solution,* $\mathrm{cl}(C - C) = Y$ *and there exists* $x' \in S$ *such that* $g(x') \in -\mathrm{qri}(C)$. *If* $(0,0) \notin \mathrm{qri}(\mathcal{E}_{v(P^C)})$, *then* $v(P^C) = v(D^{C_L})$ *and* (D^{C_L}) *has an optimal solution.*

Remark 22.7. Corollary 22.6 improves [60, Theorem 4], where for strong Lagrange duality along with the assumptions $\mathrm{cl}(C - C) = Z$ and that there exists $x' \in S$ such that $g(x') \in -\mathrm{qri}(C)$ a further condition, called *Assumption S*, has been considered. Actually, this condition turns out to be not only sufficient, but also necessary for strong duality, making the other hypotheses superfluous. For more details on this topic we refer to [17].

Chapter VI
Applications of the Duality to Monotone Operators

23 Monotone Operators and Their Representative Functions

The theory of monotone operators captured the attention of mathematicians not only because of the fineness of the results but also because of the large number of applications, especially in fields like nonlinear analysis, variational inequalities and partial differential equations (see for instance [88, 130]). Different attempts to establish links to the convex analysis have been made (see [90, 91, 119]), but the most fruitful ones turned out to be the ones based on the so-called *Fitzpatrick function* discovered by Simons Fitzpatrick in [70]. Neglected for many years until re-popularized in [7, 8, 52, 98, 104–106, 121], this class of functions along with its extensions have given rise to a great number of publications which rediscovered and extended the important results of the theory of monotone operators by using tools from the convex analysis. The investigations we make in this chapter are to be seen belonging to this class of results, whereby, we concentrate ourselves on results based on the conjugate duality theory.

Let X be a nonzero Banach space, X^* its topological dual space and X^{**} its bidual space, while by $\langle \cdot, \cdot \rangle$ we denote the duality products in both $X^* \times X$ and $X^{**} \times X^*$. A set-valued operator $S : X \rightrightarrows X^*$ is said to be *monotone* if

$$\langle y^* - x^*, y - x \rangle \geq 0 \text{ whenever } x^* \in S(x) \text{ and } y^* \in S(y).$$

For a set-valued operator $S : X \rightrightarrows X^*$ we denote its *graph* by $G(S) = \{(x, x^*) \in X \times X^* : x^* \in S(x)\}$, its *domain* by $D(S) = \{x \in X : S(x) \neq \emptyset\}$ and its *range* by $R(S) = \cup_{x \in X} S(x)$.

The monotone operator S is called *maximal monotone* if its graph is not properly contained in the graph of any other monotone operator $S' : X \rightrightarrows X^*$. The classical example of a maximal monotone operator is the subdifferential of a proper, convex and lower semicontinuous function (see [114]). Thus denoting by $\|\cdot\|$ and $\|\cdot\|_*$ the norm on X and X^*, respectively, the *duality map* $J : X \rightrightarrows X^*$, defined as

$$J(x) = \partial\left(\frac{1}{2}\|\cdot\|^2\right)(x) = \left\{x^* \in X^* : \|x\|^2 = \|x^*\|_*^2 = \langle x^*, x \rangle\right\}, x \in X$$

R.I. Boţ, *Conjugate Duality in Convex Optimization*, Lecture Notes in Economics and Mathematical Systems 637, DOI 10.1007/978-3-642-04900-2_6,

is a maximal monotone operator. However, there exist maximal monotone operators which are not subdifferentials (see [119]).

An element $(u, u^*) \in X \times X^*$ is said to be *monotonically related* to the graph of S if

$$\langle y^* - u^*, y - u\rangle \geq 0 \text{ for all } (y, y^*) \in G(S).$$

A monotone operator S is maximal monotone if and only if the set of monotonically related elements to $G(S)$ is exactly $G(S)$.

To any monotone operator $S : X \rightrightarrows X^*$, we associate the *Fitzpatrick function* $\varphi_S : X \times X^* \to \overline{\mathbb{R}}$ defined by

$$\varphi_S(x, x^*) = \sup\{\langle y^*, x\rangle + \langle x^*, y\rangle - \langle y^*, y\rangle : y^* \in S(y)\},$$

which is obviously convex and weak-weak* lower semicontinuous. It was introduced by Fitzpatrick in [70] and it proved to play an important role in the theory of maximal monotone operators, revealing some connections to the convex analysis. Consider also the *coupling function* $c : X \times X^* \to \mathbb{R}$, $c(x, x^*) = \langle x^*, x\rangle$, and $c_S : X \times X^* \to \overline{\mathbb{R}}$, $c_S = c + \delta_{G(S)}$. The function $\psi_S = \operatorname{cl}_{\|\cdot\| \times \|\cdot\|_*}(\operatorname{co} c_S)$ is the so-called *Penot function* (see [104, 105] for properties of this function in reflexive Banach spaces) and it is well-linked to the function of Fitzpatrick.

Having a function $f : A \times B \to \overline{\mathbb{R}}$, where A and B are nonempty sets, we denote by $f^\top$ the *transpose* of f, namely the function $f^\top : B \times A \to \overline{\mathbb{R}}$, $f^\top(b, a) = f(a, b)$. If $h : X \times X^* \to \overline{\mathbb{R}}$ is a given function we denote by $\widehat{h^*} : X^* \times X^{**} \to \overline{\mathbb{R}}$ its conjugate function and by $h^* : X^* \times X \to \overline{\mathbb{R}}$, $h^*(x^*, x) = \widehat{h^*}(x^*, \widehat{x})$, its *canonical embedding* to $X^* \times X$. Then one has on $X \times X^*$ that $\psi_S^{*\top} = \varphi_S$. The following result is due to Fitzpatrick.

Proposition 23.1 (cf. [70]). *Let $S : X \rightrightarrows X^*$ be a maximal monotone operator. Then*

(i) $\varphi_S(x, x^*) \geq \langle x^*, x\rangle$ for all $(x, x^*) \in X \times X^*$;
(ii) $G(S) = \{(x, x^*) \in X \times X^* : \varphi_S(x, x^*) = \langle x^*, x\rangle\}$.

Motivated by these properties of the Fitzpatrick function, the notion of *representative function* of a monotone operator was introduced and studied in the literature.

Definition 23.2 (cf. [70]). For $S : X \rightrightarrows X^*$ a monotone operator we call *representative function* of S a convex and strong lower semicontinuous function $h_S : X \times X^* \to \overline{\mathbb{R}}$ fulfilling

$$h_S \geq c \text{ and } G(S) \subseteq \{(x, x^*) \in X \times X^* : h_S(x, x^*) = \langle x^*, x\rangle\}.$$

If $G(S) \neq \emptyset$ (in particular if S is maximal monotone), then every representative function of S is proper. From Proposition 23.1 it follows that the Fitzpatrick function associated to a maximal monotone operator is a representative function of the operator. The next result is a direct consequence of the ones from [51] (see also [96, Proposition 1.2, Theorem 4.2(1)]).

Proposition 23.3. *Let $S : X \rightrightarrows X^*$ be a maximal monotone operator and h_S be a representative function of S. Then:*

(i) $\varphi_S(x,x^*) \le h_S(x,x^*) \le \psi_S(x,x^*)$ *for all* $(x,x^*) \in X \times X^*$;
(ii) the function $h_S^{*\top}$ *is a representative function of* S;
(iii) $G(S) = \{(x,x^*) \in X \times X^* : h_S(x,x^*) = \langle x^*,x\rangle\} = \{(x,x^*) \in X \times X^* : h_S^{*\top}(x,x^*) = \langle x^*,x\rangle\}$.

Viceversa, if S is maximal monotone and $h_S : X \times X^* \to \overline{\mathbb{R}}$ is a convex and strong lower semicontinuous function that fulfills on $X \times X^*$ $\varphi_S \le h_S \le \psi_S$, then h_S is a representative function of S.

If $f : X \to \overline{\mathbb{R}}$ is a proper, convex and lower semicontinuous function, then a representative function of the maximal monotone operator $\partial f : X \rightrightarrows X^*$ is the function $(x,x^*) \mapsto f(x) + f^*(x^*)$. Indeed, for all $(x,x^*) \in X \times X^*$ it holds $f(x) + f^*(x^*) \ge \langle x^*,x\rangle$. Moreover, $(x,x^*) \in G(\partial f)$ if and only if $f(x) + f^*(x^*) = \langle x^*,x\rangle$.

If f is a sublinear and lower semicontinuous function, then the operator $\partial f : X \rightrightarrows X^*$ has a unique representative function, namely the function $(x,x^*) \mapsto f(x) + f^*(x^*)$ (see [44, Theorem 3.1] and [106, Example 3]).

If X is a Hilbert space, then there exists a unique representative function of the maximal monotone operator $\partial(\delta_C) : X \rightrightarrows X$, where C is a nonempty convex and closed subset of X. Indeed, by [6, Example 3.1], the Fitzpatrick function of $\partial(\delta_C) = N_C$ is $\varphi_{\partial(\delta_C)}(x,x^*) = \delta_C(x) + \sigma_C(x^*), (x,x^*) \in X \times X^*$. This implies that $\varphi^*_{\partial(\delta_C)} = \varphi^{*\top}_{\partial(\delta_C)} = \varphi_{\partial(\delta_C)}$ for all $(x,x^*) \in X \times X^*$. Thus the unique representative function is $(x,x^*) \mapsto \delta_C(x) + \sigma_C(x^*)$.

In the remaining of this section, we give some characterizations for the maximality of a monotone operator in reflexive Banach spaces. Under this assumptions we identify the dual of $X \times X^*$ with $X^* \times X$ and consider as duality product

$$\langle (y^*,y),(x,x^*)\rangle = \langle y^*,x\rangle + \langle x^*,y\rangle$$

for $(x,x^*) \in X \times X^*$ and $(y^*,y) \in X^* \times X$. Remaining consistent with the notations above, the conjugate of a function $f : X \times X^* \to \overline{\mathbb{R}}$ will be denoted by $f^* : X^* \times X \to \overline{\mathbb{R}}$, $f^*(y^*,y) = \sup_{(x,x^*)\in X\times X^*}\{\langle y^*,x\rangle + \langle x^*,y\rangle - f(x,x^*)\}$.

Theorem 23.4 (cf. [119]). *A monotone operator S on a reflexive Banach space X is maximal if and only if $G(S) + G(-J) = X \times X^*$ or, equivalently, the mapping $S(x + \cdot) + J(\cdot)$ is surjective for all $x \in X$.*

One should notice that the result above has been given in [119] under the name *"$-J$" criterion for maximality*. Another important characterization of the maximality in reflexive Banach spaces was formulated by Burachik and Svaiter in [52, Theorem 3.1] and by Penot and Zălinescu in [108, Proposition 2.1].

Theorem 23.5 (cf. [52, 108]). *Let X be a reflexive Banach space. For any proper, convex and lower semicontinuous function $f : X \times X^* \to \overline{\mathbb{R}}$ with $f \ge c$ one has that the operator having as graph $\{(x,x^*) \in X \times X^* : f(x,x^*) = \langle x^*,x\rangle\}$ is maximal monotone if and only if $f^* \ge c^\top$.*

24 Maximal Monotonicity of the Operator $S + A^*TA$

Consider X and Y reflexive Banach spaces, $A : X \to Y$ a linear continuous operator and $S : X \rightrightarrows X^*$ and $T : Y \rightrightarrows Y^*$ two maximal monotone operators with representative functions h_S and h_T, respectively, such that $A(\Pr_X(\operatorname{dom} h_S)) \cap \Pr_Y(\operatorname{dom} h_T) \neq \emptyset$. The operator $S + A^*TA : X \rightrightarrows X^*$ defined by $(S + A^*TA)(x) = S(x) + (A^* \circ T \circ A)(x)$ is monotone, but not always maximal monotone. In this section we prove, using an idea due to Borwein (cf. [7]), that this operator is maximal monotone, provided a closedness-type regularity condition is fulfilled. To this aim, we use the following result, which is a direct consequence of Theorem 9.1 (for more details, see [23]).

Theorem 24.1. *Let X, Y and U be separated locally convex spaces, $g : Y \to \overline{\mathbb{R}}$ a proper, convex and lower semicontinuous function and $A : X \to Y$ a linear continuous operator such that $R(A) \cap \operatorname{dom} g \neq \emptyset$. Consider moreover the linear operator $M : U \to X^*$ and let τ be any locally convex topology on X^* giving X as dual. Then following statements are equivalent:*

(i) $A^* \times \mathrm{id}_{\mathbb{R}}(\operatorname{epi} g^*)$ *is closed regarding the set* $R(M) \times \mathbb{R}$ *in* $(X^*, \tau) \times \mathbb{R}$*;*
(ii) $(g \circ A)^*(Mu) = \min\{g^*(y^*) : A^*y^* = Mu\}\ \forall u \in U$.

Let us notice that one can take in the preceding proposition as τ the weak* topology on X^* or the strong topology on X^* in case X is a reflexive Banach space.

Next we prove that in this case the following regularity condition

$$(RC^{S+A^*TA})\ \{(x^* + A^*y^*, x, y, r) : h_S^*(x^*, x) + h_T^*(y^*, y) \leq r\} \text{ is closed regarding the set } X^* \times \Delta_X^A \times \mathbb{R},$$

where $\Delta_X^A = \{(x, Ax) : x \in X\}$, is sufficient for the maximality of the operator $S + A^*TA$ (cf. [18]).

Theorem 24.2. *If (RC^{S+A^*TA}) is fulfilled, then $S + A^*TA$ is a maximal monotone operator.*

Proof. Consider $z \in X$ and $z^* \in X^*$ some fixed elements. The result will follow as a consequence of Theorem 23.4, more precisely we prove that there exists $\bar{x} \in X$ such that $z^* \in (S + A^*TA)(\bar{x} + z) + J(\bar{x})$. Consider the functions $F, G : X \times X^* \to \overline{\mathbb{R}}$ defined by

$$F(x, x^*) = \inf_{\substack{(u^*, y^*) \in X^* \times Y^* \\ u^* + A^*y^* = x^* + z^*}} \{h_S(x+z, u^*) + h_T(A(x+z), y^*) - \langle u^*, z\rangle - \langle y^*, Az\rangle\}$$

and

$$G(x, x^*) = \frac{1}{2}\|x\|^2 + \frac{1}{2}\|x^*\|_*^2 - \langle z^*, x\rangle.$$

First we compute the conjugates of F and G. For $(\omega^*, \omega) \in X^* \times X$ we have

$$\begin{aligned} F^*(\omega^*, \omega) &= \sup_{(x,x^*)\in X\times X^*} \Big\{ \langle \omega^*, x\rangle + \langle x^*, \omega\rangle - \inf_{\substack{(u^*,y^*)\in X^*\times Y^* \\ u^*+A^*y^*=x^*+z^*}} \{h_S(x+z, u^*) \\ &\quad + h_T(A(x+z), y^*) - \langle u^*, z\rangle - \langle y^*, Az\rangle\}\Big\} \\ &= \sup_{\substack{x\in X \\ (u^*,y^*)\in X^*\times Y^*}} \{\langle \omega^*, x\rangle + \langle u^* + A^*y^* - z^*, \omega\rangle \\ &\quad - h_S(x+z, u^*) - h_T(A(x+z), y^*) + \langle u^*, z\rangle + \langle y^*, Az\rangle\} \\ &= \sup_{\substack{x\in X \\ (u^*,y^*)\in X^*\times Y^*}} \{\langle \omega^*, u\rangle + \langle u^*, \omega + z\rangle + \langle y^*, A(\omega+z)\rangle \\ &\quad - h_S(u, u^*) - h_T\big(Au, y^*\big)\} - \langle \omega^*, z\rangle - \langle z^*, \omega\rangle. \end{aligned}$$

Considering the functions $g : X\times Y\times X^*\times Y^* \to \overline{\mathbb{R}}$, $g(x, y, x^*, y^*) = h_S(x, x^*) + h_T(y, y^*)$, $B : X\times X^*\times Y^* \to X\times Y\times X^*\times Y^*$, $B(x, x^*, y^*) = (x, Ax, x^*, y^*)$ and $M : X \times X^* \to X^* \times X \times Y$, $M(\omega, \omega^*) = (\omega^*, \omega, A\omega)$, we have that

$$F^*(\omega^*, \omega) = (g \circ B)^* \big(M(\omega + z, \omega^*)\big) - \langle \omega^*, z\rangle - \langle z^*, \omega\rangle.$$

One can easily check that the conjugate of g is $g^* : X^* \times Y^* \times X \times Y \to \overline{\mathbb{R}}$, $g^*(x^*, y^*, x, y) = h_S^*(x^*, x) + h_T^*(y^*, y)$ and the adjoint operator of the linear continuous mapping B is $B^* : X^*\times Y^*\times X\times Y \to X^*\times X\times Y$, $B^*(x^*, y^*, x, y) = (x^* + A^*y^*, x, y)$. Hence $B^* \times \mathrm{id}_{\mathbb{R}}(\mathrm{epi}\, g^*) = \{(B^*(x^*, y^*, x, y), r) : g^*(x^*, y^*, x, y) \le r\} = \big\{(x^* + A^*y^*, x, y, r) : h_S^*(x^*, x) + h_T^*(y^*, y) \le r\big\}$ and because of $R(M) \times \mathbb{R} = X^* \times \Delta_X^A \times \mathbb{R}$, the regularity condition (RC^{S+A^*TA}) is nothing else than $B^* \times \mathrm{id}_{\mathbb{R}}(\mathrm{epi}\, g^*)$ is closed regarding the set $R(M) \times \mathbb{R}$. By Theorem 24.1 we have that $(g \circ B)^* (M(\omega + z, \omega^*)) = \min\{g^*(a^*, b^*, a, b) : B^*(a^*, b^*, a, b) = M(\omega + z, \omega^*)\}$ (the fact that $R(B) \cap \mathrm{dom}\, g \neq \emptyset$ follows from the assumption $A(\mathrm{Pr}_X(\mathrm{dom}\, h_S)) \cap \mathrm{Pr}_Y(\mathrm{dom}\, h_T) \neq \emptyset$). Thus

$$\begin{aligned} F^*(\omega^*, \omega) &= \min_{(a^*+A^*b^*,a,b)=(\omega^*,\omega+z,A(\omega+z))} \big\{h_S^*(a^*, a) + h_T^*(b^*, b)\big\} \\ &\quad - \langle \omega^*, z\rangle - \langle z^*, \omega\rangle. \end{aligned}$$

For the conjugate of G we have

$$\begin{aligned} G^*(\omega^*, \omega) &= \sup_{\substack{x\in X \\ x^*\in X^*}} \left\{ \langle \omega^*, x\rangle + \langle x^*, \omega\rangle - \frac{1}{2}\|x\|^2 - \frac{1}{2}\|x^*\|_*^2 + \langle z^*, x\rangle \right\} \\ &= \frac{1}{2}\|\omega^* + z^*\|_*^2 + \frac{1}{2}\|\omega\|^2. \end{aligned}$$

For every $(x, x^*) \in X \times X^*$ and $(u^*, y^*) \in X^* \times Y^*$ such that $u^* + A^* y^* = x^* + z^*$ we obtain by the definition of the representative function that

$$\begin{aligned} &h_S(x+z, u^*) + h_T(A(x+z), y^*) - \langle u^*, z\rangle - \langle y^*, Az\rangle + G(x, x^*) \\ &\geq \langle u^*, x+z\rangle + \langle y^*, A(x+z)\rangle - \langle u^*, z\rangle - \langle y^*, Az\rangle + G(x, x^*) \\ &= \langle u^*, x\rangle + \langle A^* y^*, x\rangle + G(x, x^*) = \langle u^*, x\rangle + \langle x^* + z^* - u^*, x\rangle + G(x, x^*) \\ &= \langle x^* + z^*, x\rangle + G(x, x^*) = \frac{1}{2}\|x\|^2 + \frac{1}{2}\|x^*\|_*^2 + \langle x^*, x\rangle \geq 0. \end{aligned}$$

This implies $F(x, x^*) + G(x, x^*) \geq 0$ for all $(x, x^*) \in X \times X^*$, that is

$$\inf_{(x,x^*) \in X \times X^*} \{F(x, x^*) + G(x, x^*)\} \geq 0.$$

The functions F and G are convex and the latter is continuous. Thus by Fenchel's Duality Theorem, Theorem 2.2 ((RC_1^{id}) is verified), there exists a pair $(\bar{x}^*, \bar{x})$ such that

$$\begin{aligned} \inf_{(x,x^*) \in X \times X^*} \{F(x, x^*) + G(x, x^*)\} &= \max_{(x^*,x) \in X^* \times X} \{-F^*(x^*, x) - G^*(-x^*, -x)\} \\ &= -F^*(\bar{x}^*, \bar{x}) - G^*(-\bar{x}^*, -\bar{x}). \end{aligned}$$

Therefore there exists $(\bar{a}^*, \bar{a}, \bar{b}^*, \bar{b}) \in X^* \times X \times Y^* \times Y$ such that $(\bar{a}^* + A^*\bar{b}^*, \bar{a}, \bar{b}) = (\bar{x}^*, \bar{x} + z, A(\bar{x} + z))$ and

$$h_S^*(\bar{a}^*, \bar{a}) + h_T^*(\bar{b}^*, \bar{b}) - \langle \bar{x}^*, z\rangle - \langle z^*, \bar{x}\rangle + \frac{1}{2}\| - \bar{x}^* + z^*\|_*^2 + \frac{1}{2}\| - \bar{x}\|^2 \leq 0.$$

Taking into account that $\bar{b} = A(\bar{x} + z) = A\bar{a}$, $\bar{x} = \bar{a} - z$ and $\bar{x}^* = \bar{a}^* + A^*\bar{b}^*$, we obtain

$$\begin{aligned} 0 &\geq \left(h_S^*(\bar{a}^*, \bar{a}) - \langle \bar{a}^*, \bar{a}\rangle\right) + \left(h_T^*(\bar{b}^*, \bar{b}) - \langle \bar{b}^*, \bar{b}\rangle\right) + \langle \bar{a}^*, \bar{a}\rangle + \langle \bar{b}^*, \bar{b}\rangle \\ &-\langle \bar{x}^*, z\rangle - \langle z^*, \bar{x}\rangle + \frac{1}{2}\| - \bar{x}^* + z^*\|_*^2 + \frac{1}{2}\| - \bar{x}\|^2 = \left(h_S^*(\bar{a}^*, \bar{a}) - \langle \bar{a}^*, \bar{a}\rangle\right) \\ &+ \left(h_T^*(\bar{b}^*, \bar{b}) - \langle \bar{b}^*, \bar{b}\rangle\right) + \langle \bar{a}^*, \bar{a}\rangle + \langle \bar{b}^*, A\bar{a}\rangle - \langle \bar{a}^*, z\rangle - \langle \bar{b}^*, Az\rangle - \langle z^*, \bar{a}\rangle + \langle z^*, z\rangle \\ &+\frac{1}{2}\| - \bar{a}^* - A^*\bar{b}^* + z^*\|_*^2 + \frac{1}{2}\| - \bar{a} + z\|^2 = \left(h_S^*(\bar{a}^*, \bar{a}) - \langle \bar{a}^*, \bar{a}\rangle\right) + \left(h_T^*(\bar{b}^*, \bar{b}) - \langle \bar{b}^*, \bar{b}\rangle\right) \\ &+\frac{1}{2}\| - \bar{a}^* - A^*\bar{b}^* + z^*\|_*^2 + \frac{1}{2}\| - \bar{a} + z\|^2 \\ &+\langle -\bar{a}^* - A^*\bar{b}^* + z^*, -\bar{a} + z\rangle \geq 0, \end{aligned}$$

where the last inequality follows by using Proposition 23.3. Hence the inequalities above must be fulfilled as equalities, that is

$$h_S^*(\bar{a}^*, \bar{a}) = \langle \bar{a}^*, \bar{a}\rangle, \ h_T^*(\bar{b}^*, \bar{b}) = \langle \bar{b}^*, \bar{b}\rangle$$

and

$$\frac{1}{2}\| -\bar{a}^* - A^*\bar{b}^* + z^*\|_*^2 + \frac{1}{2}\| -\bar{a} + z\|^2 + \langle -\bar{a}^* - A^*\bar{b}^* + z^*, -\bar{a} + z\rangle = 0.$$

Using again Proposition 23.3 and the definition of the duality map J, we equivalently obtain that $\bar{a}^* \in S(\bar{a})$, $\bar{b}^* \in T(\bar{b})$ and $z^* - \bar{a}^* - A^*\bar{b}^* \in J(\bar{a} - z)$. Employing once more the relation $(\bar{a}^* + A^*\bar{b}^*, \bar{a}, \bar{b}) = (\bar{x}^*, \bar{x} + z, A(\bar{x} + z))$ we get $z^* - \bar{x}^* \in J(\bar{x})$ and $\bar{x}^* = \bar{a}^* + A^*\bar{b}^* \in S(\bar{a}) + A^*T(A\bar{a}) = (S + A^*TA)(\bar{a}) = (S + A^*TA)(\bar{x} + z)$. Finally we have

$$z^* = \bar{x}^* + (z^* - \bar{x}^*) \in (S + A^*TA)(\bar{x} + z) + J(\bar{x}).$$

As z and z^* have been arbitrary chosen, Theorem 23.4 yields the conclusion. □

Remark 24.3. One should notice that the maximal monotonicity for $S + A^*TA$ is valid even if asking only that

$$(\widetilde{RC}^{S+A^*TA})\ \{(x^* + A^*y^*, x, y, r) : h_S^*(x^*, x) + h_T^*(y^*, y) \leq r\} \text{ is closed.}$$

The regularity conditions for the maximality of $S + A^*TA$ that one can find in the literature are of *generalized interior point-type*. We mention here only the weakest conditions in this sense, namely the one given by Pennanen in [103]

$$0 \in \operatorname{ri}(A(D(S)) - D(T)),$$

which can be equivalently formulated as

$$0 \in \operatorname{ri}(A(\operatorname{co}(D(S))) - \operatorname{co}(D(T))$$

and, respectively, the one given by Penot and Zălinescu in [108]

$$(RC_{PZ}^{S+A^*TA})\ 0 \in \operatorname{sqri}\left(A(\operatorname{Pr}_X(\operatorname{dom} h_S)) - \operatorname{Pr}_Y(\operatorname{dom} h_T)\right),$$

with the equivalent formulations

$$0 \in \operatorname{sqri}(A(D(S)) - D(T))$$

and

$$0 \in \operatorname{sqri}(A(\operatorname{co}(D(S))) - \operatorname{co}(D(T)).$$

Here for $U \subseteq X$ by $\operatorname{ri}(U)$ we denote the set which is equal to the interior of U with respect to its affine hull in case the latter is closed, being the empty set, otherwise. According to [108, Corollary 3.6], one has

$$\operatorname{ri}(A(\operatorname{co}(D(S))) - \operatorname{co}(D(T)) = \operatorname{sqri}(A(\operatorname{co}(D(S))) - \operatorname{co}(D(T))$$

and this means that all the generalized interior point conditions above are equivalent. Next we show that $(\widetilde{RC}^{S+A^*TA})$ and, consequently, (RC^{S+A^*TA}) are implied by the aforementioned conditions.

We assume that $(RC_{PZ}^{S+A^*TA})$ is fulfilled and consider the following functions $s : X \times X^* \times Y^* \to \overline{\mathbb{R}}, s(x, x^* y^*) = h_S(x, x^*), t : Y \times X^* \times Y^* \to \overline{\mathbb{R}}, t(y, x^*, y^*) = h_T(y, y^*)$ and $C : X \times X^* \times Y^* \to Y \times X^* \times Y^*, C(x, x^*, y^*) = (Ax, x^*, y^*)$. One can see by direct computation that

$$s^* : X^* \times X \times Y \to \overline{\mathbb{R}},\ s^*(x^*, x, y) = \begin{cases} h_S^*(x^*, x), & \text{if } y = 0, \\ +\infty, & \text{otherwise,} \end{cases}$$

$$t^* : Y^* \times X \times Y \to \overline{\mathbb{R}},\ t^*(y^*, x, y) = \begin{cases} h_T^*(y^*, y), & \text{if } x = 0, \\ +\infty, & \text{otherwise} \end{cases}$$

and $C^* : Y^* \times X \times Y \to X^* \times X \times Y, C^*(y^*, x, y) = (A^*y^*, x, y)$. As

$$\operatorname{cone}(A(\Pr_X(\operatorname{dom} h_S)) - \Pr_Y(\operatorname{dom} h_T)) \times X^* \times Y^* = \operatorname{cone}(C(\operatorname{dom} s) - \operatorname{dom} t)$$

it follows that $0 \in \operatorname{sqri}(C(\operatorname{dom} s) - \operatorname{dom} t)$. Thus the regularity condition (RC_2^A) introduced in Section 2 is fulfilled and, by Theorem 5.5, one has for all $(x^*, x, y) \in X^* \times X \times Y$ that

$$\begin{aligned}(s + t \circ C)^*(x^*, x, y) &= \min_{(u^*, v, z) \in Y^* \times X \times Y} \left\{ s^*\left((x^*, x, y) - C^*(u^*, v, z)\right) \right. \\ &\quad \left. + t^*(u^*, v, z) \right\}.\end{aligned}$$

On the other hand, by Theorem 7.3, this is the same as assuming that $\operatorname{epi} s^* + C^* \times \operatorname{id}_{\mathbb{R}}(\operatorname{epi} t^*)$ is closed in $X^* \times X \times Y \times \mathbb{R}$. As $\operatorname{epi} s^* = \{(x^*, x, 0, r) : h_S^*(x^*, x) \leq r\}$ and $C^* \times \operatorname{id}_{\mathbb{R}}(\operatorname{epi} t^*) = \{(A^*y^*, 0, y, r) : h_T^*(y^*, y) \leq r\}$ we have in conclusion that $\{(x^* + A^*y^*, x, y, r) : h_S^*(x^*, x) + h_T^*(y^*, y) \leq r\}$ is closed and hence closed regarding the set $X^* \times \Delta_X^A \times \mathbb{R}$. In other words, (RC^{S+A^*TA}) is fulfilled. Actually we proved that $(RC_{PZ}^{S+A^*TA}) \Rightarrow (\widetilde{RC}^{S+A^*TA}) \Rightarrow (RC^{S+A^*TA})$. The fact that (RC^{S+A^*TA}) is indeed weaker than $(RC_{PZ}^{S+A^*TA})$ is shown by an example in the next section.

Remark 24.4. Consider $f : X \to \overline{\mathbb{R}}$ and $g : Y \to \overline{\mathbb{R}}$ two proper, convex and lower semicontinuous functions such that $A(\operatorname{dom} f) \cap \operatorname{dom} g \neq \emptyset$ and the maximal monotone operators $\partial f : X \rightrightarrows X^*$ and $\partial g : Y \rightrightarrows Y^*$ along with their representative functions $(x, x^*) \mapsto f(x) + f^*(x^*)$ and $(y, y^*) \mapsto g(y) + g^*(y^*)$, respectively. By Theorem 24.2 it yields that if

$$\{(x^* + A^*y^*, x, y, r) : f(x) + f^*(x^*) + g(y) + g^*(y^*) \leq r\} \text{ is closed regarding the set } X^* \times \Delta_X^A \times \mathbb{R}, \tag{24.1}$$

then $\partial f + A^*\partial g A$ is a maximal monotone operator. As one always has that $\partial(f + g \circ A) \supseteq \partial f + A^*\partial g A$ and $\partial(f + g \circ A)$ is monotone, this is equivalent to $\partial(f + g \circ A) = \partial f + A^*\partial g A$.

Next we prove that (24.1) is equivalent to

$$\text{epi } f^* + A^* \times \text{id}_{\mathbb{R}}(\text{epi } g^*) \text{ is closed in } X^* \times \mathbb{R}. \tag{24.2}$$

To this end we make the notations

$$U := \{(x^* + A^*y^*, x, y, r) : f(x) + f^*(x^*) + g(y) + g^*(y^*) \leq r\}$$

and $V := X^* \times \Delta_X^A \times \mathbb{R}$. Hence (24.1) can be written as $\text{cl}(U) \cap V = U \cap V$.

(24.1)⇒(24.2) If (u^*, r) is an element in $\text{cl}(\text{epi } f^* + A^* \times \text{id}_{\mathbb{R}}(\text{epi } g^*))$, then there exist the sequences $\{x_n^* : n \in \mathbb{N}\} \subseteq X^*$, $\{y_n^* : n \in \mathbb{N}\} \subseteq Y^*$, $\{r_n : n \in \mathbb{N}\} \subseteq \mathbb{R}$ and $\{t_n : n \in \mathbb{N}\} \subseteq \mathbb{R}$ such that

$$\begin{cases} f^*(x_n^*) \leq r_n, g^*(y_n^*) \leq t_n \ \forall n \in \mathbb{N}, \\ x_n^* + A^*y_n^* \to u^*, r_n + t_n \to r \ (n \to +\infty). \end{cases}$$

By the hypothesis, there exists $x' \in \text{dom } f \cap A^{-1}(\text{dom } g)$. We get that for all $n \in \mathbb{N}$ $(x_n^* + A^*y_n^*, x', Ax', r_n + t_n + f(x') + g(Ax')) \in U$, which implies that $(u^*, x', Ax', r + f(x') + g(Ax')) \in \text{cl}(U) \cap V = U \cap V$. Thus there exist $x^* \in X^*$ and $y^* \in Y^*$ such that $u^* = x^* + A^*y^*$ and $f(x') + f^*(x^*) + g(Ax') + g^*(y^*) \leq r + f(x') + g(Ax')$, that is $f^*(x^*) + g^*(y^*) \leq r$. Consequently,

$$(u^*, r) = (x^*, r - g^*(y^*)) + (A^*y^*, g^*(y^*)) \in \text{epi } f^* + A^* \times \text{id}_{\mathbb{R}}(\text{epi } g^*)$$

and so (24.2) follows.

(24.2)⇒(24.1) Let $(z^*, z, Az, r) \in \text{cl}(U) \cap V$ be fixed. There exist some sequences $\{x_n^* : n \in \mathbb{N}\} \subseteq X^*$, $\{y_n^* : n \in \mathbb{N}\} \subseteq Y^*$, $\{x_n : n \in \mathbb{N}\} \subseteq X$, $\{y_n : n \in \mathbb{N}\} \subseteq Y$ and $\{r_n : n \in \mathbb{N}\} \subseteq \mathbb{R}$ such that

$$\begin{cases} f(x_n) + f^*(x_n^*) + g(y_n) + g^*(y_n^*) \leq r_n \ \forall n \in \mathbb{N}, \\ x_n^* + A^*y_n^* \to z^*, x_n \to z, y_n \to Az, r_n \to r \ (n \to +\infty). \end{cases}$$

For all $n \in \mathbb{N}$ we have $(f + g \circ A)^*(x_n^* + A^*y_n^*) \leq f^*(x_n^*) + g^*(y_n^*)$ and from here $f(x_n) + g(y_n) + (f + g \circ A)^*(x_n^* + A^*y_n^*) \leq r_n$. The lower semicontinuity of the functions involved yields $f(z) + g(Az) + (f + g \circ A)^*(z^*) \leq r$. Thus, by Theorem 7.2, $(z^*, r - f(z) - g(Az)) \in \text{epi}(f + g \circ A)^* = \text{cl}(\text{epi } f^* + A^* \times \text{id}_{\mathbb{R}}(\text{epi } g^*)) = \text{epi } f^* + A^* \times \text{id}_{\mathbb{R}}(\text{epi } g^*)$. Hence there exist $x^* \in X^*$, $y^* \in Y^*$ and $s, t \in \mathbb{R}$ such that $(z^*, r - f(z) - g(Az)) = (x^*, s) + (A^*y^*, t)$, $f^*(x^*) \leq s$ and $g^*(y^*) \leq t$. Therefore $f^*(x^*) + g^*(y^*) \leq s + t = r - f(z) - g(Az) \Leftrightarrow f(z) + f^*(x^*) + g(Az) + g^*(y^*) \leq r$. This means that $(z^*, z, Az, r) = (x^* + A^*y^*, z, Az, r) \in U \cap V$. Consequently, $\text{cl}(U) \cap V \subseteq U \cap V$ and (24.1) is fulfilled.

The fact that the closedness-type condition in (24.2) is sufficient for having the subdifferential sum formula follows as consequence of Theorem 7.5, even when working in separated locally convex spaces.

25 The Maximality of A^*TA and $S + T$

In the following we deal with some particular cases of Theorem 24.2 and formulate closedness-type sufficient conditions for the maximality of the composition of a maximal monotone operator with a linear continuous operator and of the sum of two maximal monotone operators, respectively. By some examples, we illustrate that these conditions are indeed weaker than the generalized interior point ones given until now in the literature.

As a first special case of Theorem 24.2, we take $S : X \rightrightarrows X^*$, $S(x) = \{0\}$ for all $x \in X$. Thus $G(S) = X \times \{0\}$ and $S + A^*TA = A^*TA$. We show that the operator S has an unique representative function. To this end, we compute first its Fitzpatrick function φ_S. We have for all $(x, x^*) \in X \times X^*$

$$\varphi_S(x, x^*) = \sup\{\langle y^*, x\rangle + \langle x^*, y\rangle - \langle y^*, y\rangle : y^* \in S(y)\}$$

$$= \begin{cases} 0, & \text{if } x^* = 0, \\ +\infty, & \text{otherwise} \end{cases} = \delta_{X\times\{0\}}.$$

If h_S is a representative function of S, then by Proposition 23.3 we get $\{(x, x^*) : h_S(x, x^*) = \langle x^*, x\rangle\} = G(S) = X \times \{0\}$, hence $h_S(x, 0) = 0$ for all $x \in X$. Let $x \in X$ and $x^* \in X^* \setminus \{0\}$ be arbitrary elements. Since $h_S(x, x^*) \geq \varphi_S(x, x^*) = +\infty$, it follows that $h_S = \delta_{X\times\{0\}}$. As $h_S^{*\top}$ is again a representative function of S, we obtain that $h_S^{*\top} = \delta_{X\times\{0\}}$, that is $h_S^* = \delta_{\{0\}\times X}$.

The feasibility condition $A(\Pr_X(\operatorname{dom} h_S)) \cap \Pr_Y(\operatorname{dom} h_T) \neq \emptyset$ looks like $R(A) \cap \Pr_Y(\operatorname{dom} h_T)) \neq \emptyset$ and the regularity condition (RC^{S+A^*TA}) can be written as

$$\{(A^*y^*, x, y, r) : h_T^*(y^*, y) \leq r\} \text{ is closed regarding the set } X^* \times \Delta_X^A \times \mathbb{R}. \quad (25.1)$$

We prove that (25.1) is equivalent to

$$(RC^{A^*TA})\ A^* \times \mathrm{id}_Y \times \mathrm{id}_{\mathbb{R}}(\operatorname{epi} h_T^*) \text{ is closed regarding the set } X^* \times R(A) \times \mathbb{R}.$$

Notice first that (25.1) and (RC^{A^*TA}) are nothing else than

$$\{(A^*y^*, x, y, r) : h_T^*(y^*, y) \leq r\} \cap (X^* \times \Delta_X^A \times \mathbb{R})$$

$$= \operatorname{cl}\left(\{(A^*y^*, x, y, r) : h_T^*(y^*, y) \leq r\}\right) \cap (X^* \times \Delta_X^A \times \mathbb{R})$$

and

$$\{(A^*y^*,y,r) : h_T^*(y^*,y) \le r\} \cap (X^* \times R(A) \times \mathbb{R})$$
$$= \operatorname{cl}\left(\{(A^*y^*,y,r) : h_T^*(y^*,y) \le r\}\right) \cap (X^* \times R(A) \times \mathbb{R}),$$

respectively.

(25.1)$\Rightarrow$ (RC^{A^*TA}) Take $(z^*, Az, s) \in \operatorname{cl}\left(\{(A^*y^*,y,r) : h_T^*(y^*,y) \le r\}\right) \cap (X^* \times R(A) \times \mathbb{R})$. Then there exist some sequences $\{y_n^* : n \in \mathbb{N}\} \subseteq Y^*$, $\{y_n : n \in \mathbb{N}\} \subseteq Y$ and $\{r_n : n \in \mathbb{N}\} \subseteq \mathbb{R}$ such that $A^*y_n^* \to z^*, y_n \to Az, r_n \to s$ $(n \to +\infty)$ and $h_T^*(y_n^*, y_n) \le r_n$ for all $n \in \mathbb{N}$. Thus $(A^*y_n^*, z, y_n, r_n) \to (z^*, z, Az, s)$ $(n \to +\infty)$ and so, by (25.1), $(z^*, z, Az, s) \in \operatorname{cl}\left(\{(A^*y^*,x,y,r) : h_T^*(y^*,y) \le r\}\right) \cap (X^* \times \Delta_X^A \times \mathbb{R}) = \{(A^*y^*,x,y,r) : h_T^*(y^*,y) \le r\} \cap (X^* \times \Delta_X^A \times \mathbb{R})$. This implies that for $y^* \in Y^*$, $z^* = A^*y^*$ and $h_T^*(y^*, Az) \le s$, so $(z^*, Az, s) \in \{(A^*y^*,y,r) : h_T^*(y^*,y) \le r\}$. Therefore (RC^{A^*TA}) is fulfilled.

$(RC^{A^*TA}) \Rightarrow$ (25.1) Let $(z^*, z, Az, s) \in \operatorname{cl}\left(\{(A^*y^*,x,y,r) : h_T^*(y^*,y) \le r\}\right) \cap (X^* \times \Delta_X^A \times \mathbb{R})$ be fixed. Then there exist some sequences $\{y_n^* : n \in \mathbb{N}\} \subseteq Y^*$, $\{x_n : n \in \mathbb{N}\} \subseteq X$, $\{y_n : n \in \mathbb{N}\} \subseteq Y$ and $\{r_n : n \in \mathbb{N}\} \subseteq \mathbb{R}$ such that $A^*y_n^* \to z^*, x_n \to z, y_n \to Az, r_n \to s$ $(n \to +\infty)$ and $h_T^*(y_n^*, y_n) \le r_n$ for all $n \in \mathbb{N}$. We have $(A^*y_n^*, y_n, r_n) \to (z^*, Az, s)$ $(n \to +\infty)$ and obtain that $(z^*, Az, s) \in \operatorname{cl}\left(\{(A^*y^*,y,r) : h_T^*(y^*,y) \le r\}\right) \cap (X^* \times R(A) \times \mathbb{R}) = \{(A^*y^*,y,r) : h_T^*(y^*,y) \le r\} \cap (X^* \times R(A) \times \mathbb{R})$. For $y^* \in Y^*$ we get $z^* = A^*y^*$ and $h_T^*(y^*, Az) \le s$, so $(z^*, z, Az, s) \in \{(A^*y^*,x,y,r) : h_T^*(y^*,y) \le r\} \cap (X^* \times \Delta_X^A \times \mathbb{R})$. Thus (25.1) is fulfilled.

We can state now the following result (cf. [18]).

Theorem 25.1. *If (RC^{A^*TA}) is fulfilled, then A^*TA is a maximal monotone operator.*

Remark 25.2. The regularity condition (RC^{A^*TA}) and Theorem 25.1 extend the corresponding statements given in [23] in case h_T is the Fitzpatrick function of T. In the literature, different generalized interior point regularity conditions for the maximality of A^*TA have been given. We recall here the one due to Borwein (cf. [7])

$$0 \in \operatorname{core}(R(A) - D(T)),$$

which is the same as

$$0 \in \operatorname{core}(R(A) - \operatorname{co}(D(T))),$$

and the one due to Pennanen (cf. [103])

$$0 \in \operatorname{ri}(R(A) - D(T)),$$

that is the same with

$$0 \in \operatorname{ri}(R(A) - \operatorname{co}(D(T))).$$

The latter have been proved in [108, 128] to be equivalent to

$$0 \in \operatorname{sqri}(R(A) - \operatorname{co}(D(T))$$

and also to

$$0 \in \operatorname{sqri}(R(A) - D(T))$$

and

$$(RC_{PZ}^{A^*TA})\ 0 \in \operatorname{sqri}\left(R(A) - \operatorname{Pr}_Y(\operatorname{dom} h_T)\right).$$

These are the weakest generalized interior point regularity conditions in reflexive Banach spaces for the maximality of A^*TA known so far. By Remark 24.3 it follows that whenever $(RC_{PZ}^{A^*TA})$ is fulfilled, the conditions

$$(\widetilde{RC}^{A^*TA})\ A^* \times \operatorname{id}_Y \times \operatorname{id}_{\mathbb{R}}(\operatorname{epi} h_T^*) \text{ is closed}$$

and (RC^{A^*TA}) are fulfilled, too. In other words, we have $(RC_{PZ}^{A^*TA}) \Rightarrow (\widetilde{RC}^{A^*TA})$ $\Rightarrow (RC^{A^*TA})$. In the example below, we present a situation where $(RC_{PZ}^{A^*TA})$ and, consequently, the other generalized interior point conditions fail, unlike $(\widetilde{RC}^{A^*TA})$ and (RC^{A^*TA}).

Example 25.3. (cf. [23]) Let $X = \mathbb{R}$, $Y = \mathbb{R}^2$ and $T : \mathbb{R}^2 \rightrightarrows \mathbb{R}^2$ be defined as

$$T(x,y) = \begin{cases} (-\infty, 0] \times \{0\}, & \text{if } x = 0, y < 0, \\ (-\infty, 0] \times [0, +\infty), & \text{if } x = y = 0, \\ \{x\} \times \{0\}, & \text{if } x > 0, y < 0, \\ \{x\} \times [0, +\infty), & \text{if } x > 0, y = 0, \\ \emptyset, & \text{otherwise.} \end{cases}$$

Is is not difficult to notice that, considering the proper, convex and lower semicontinuous functions $f, g : \mathbb{R} \to \overline{\mathbb{R}}$, $f(x) = (1/2)x^2 + \delta_{\mathbb{R}_+}(x)$ and $g(x) = \delta_{(-\infty,0]}(x)$, we have $T(x,y) = \partial f(x) \times \partial g(y)$ for all $(x,y) \in \mathbb{R}^2$. Thus T is a maximal monotone operator. Taking $A : \mathbb{R} \to \mathbb{R}^2$, $Ax = (x,x)$, one gets for any $x \in \mathbb{R}$

$$A^* \circ T \circ A(x) = \partial f(x) + \partial g(x) = \begin{cases} \mathbb{R}, & \text{if } x = 0, \\ \emptyset, & \text{otherwise} \end{cases}$$

and one can see that $A^* \circ T \circ A$ is maximal monotone operator, too.

We consider as a representative function for T its Fitzpatrick function φ_T. We have for all $(x, y, x^*, y^*) \in \mathbb{R}^4$

$$\varphi_T(x,y,x^*,y^*) = \begin{cases} \left(\frac{x+x^*}{2}\right)^2, & \text{if } x \geq 0, x + x^* > 0, y \leq 0, y^* \geq 0, \\ 0, & \text{if } x \geq 0, x + x^* \leq 0, y \leq 0, y^* \geq 0, \\ +\infty, & \text{otherwise} \end{cases}$$

and

$$\varphi_T^*(x^*,y^*,x,y) = \begin{cases} x^2, & \text{if } x \geq 0, x \geq x^*, y^* \geq 0, y \leq 0, \\ +\infty, & \text{otherwise.} \end{cases}$$

The epigraph of its conjugate is

$$\operatorname{epi}\varphi_T^* = \bigcup_{x\geq 0}\left((-\infty,x]\times[0,+\infty)\times\{x\}\times(-\infty,0]\times[x^2,+\infty)\right)$$

and so

$$A^*\times \mathrm{id}_{\mathbb{R}^2}\times \mathrm{id}_{\mathbb{R}}(\operatorname{epi}\varphi_T^*) = \mathbb{R}\times\bigcup_{x\geq 0}\left(\{x\}\times(-\infty,0]\times[x^2,+\infty)\right),$$

which is closed. The regularity condition $(\widetilde{RC}^{A^*TA})$ is valid and the same applies for (RC^{A^*TA}).

In order to check the validity of $(RC_{PZ}^{S+A^*TA})$ we calculate $\operatorname{cone}(R(A)-D(T))$. It is clear that $D(T)=[0,+\infty)\times(-\infty,0]$ and $R(A)=\Delta_{\mathbb{R}^2}$. We have $R(A)-D(T)=\cup_{x\in\mathbb{R}}(-\infty,x]\times[x,+\infty)$ and so $\operatorname{cone}(R(A)-D(T))=\cup_{x\in\mathbb{R}}(-\infty,x]\times[x,+\infty)=\{(x,y)\in\mathbb{R}^2 : x\leq y\}$, which is not a linear subspace, and this means that the generalized interior point condition is violated.

The second important particular case of Theorem 24.2, that we treat in this section, is obtained by taking $Y=X$ and $A=\mathrm{id}_X$. Then $A^*=\mathrm{id}_{X^*}$, $S,T: X\rightrightarrows X^*$ and $S+A^*TA=S+T$. The condition $A(\mathrm{Pr}_X(\operatorname{dom}h_S))\cap \mathrm{Pr}_X(\operatorname{dom}h_T)\neq\emptyset$ becomes $\mathrm{Pr}_X(\operatorname{dom}h_S)\cap\mathrm{Pr}_X(\operatorname{dom}h_T)\neq\emptyset$ and the regularity condition (RC^{S+A^*TA}) turns out to be

(RC^{S+T}) $\{(x^*+y^*,x,y,r) : h_S^*(x^*,x)+h_T^*(y^*,y)\leq r\}$ is closed regarding the set $X^*\times\Delta_{X^2}\times\mathbb{R}$.

Theorem 24.2 yields to the following result.

Theorem 25.4. *If* (RC^{S+T}) *is fulfilled, then* $S+T$ *is a maximal monotone operator.*

Another possibility to get (RC^{S+T}) is by deriving it from (RC^{A^*TA}) when $Y=X\times X$, $A: X\to X\times X$, $Ax=(x,x)$ and $(S,T): X\times X\rightrightarrows X^*\times X^*$, $(S,T)(x,y)=S(x)\times T(y)$. In case S and T are maximal monotone operators (S,T) is also a maximal monotone operator and it holds $A^*(S,T)A(x)=S(x)+T(x)$ for all $x\in X$. After some calculations, one can see that (RC^{A^*TA}) becomes (RC^{S+T}).

Remark 25.5. The regularity condition (RC^{S+T}) and Theorem 25.4 extend the corresponding result given in [23, 28] in case h_S and h_T are the Fitzpatrick functions of S and T, respectively. We want to mention that the most famous regularity condition ensuring the maximal monotonicity of the sum of two maximal monotone operators was introduced in the pioneering work of Rockafellar [115] and it looks like

$$D(S)\cap\operatorname{int}(D(T))\neq\emptyset.$$

Rockafellar's condition has been weakened by considering the algebraic interior instead of the topological interior to (cf. [7, 105])

$$0 \in \text{core}(D(S) - D(T)),$$

which is the same as

$$0 \in \text{core}(\text{co}(D(S)) - \text{co}(D(T))).$$

A further improvement of these regularity conditions has been furnished in [56, 103] where

$$0 \in \text{ri}(D(S) - D(T))$$

or, equivalently,

$$0 \in \text{ri}(\text{co}(D(S)) - \text{co}(D(T)))$$

has been assumed. The latter conditions have been proved in [108, 128] to be equivalent to the following ones expressed via the strong quasi-relative interior (see [3])

$$0 \in \text{sqri}(D(S) - D(T))$$

and also to

$$0 \in \text{sqri}(\text{co}(D(S) - \text{co}(D(T)))$$

and

$$(RC_{PZ}^{S+T})\ 0 \in \text{sqri}\left(\text{Pr}_X(\text{dom}\, h_S) - \text{Pr}_X(\text{dom}\, h_T)\right).$$

These are the weakest generalized interior point regularity conditions in reflexive Banach spaces known so far, which guarantee that $S + T$ is a maximal monotone operator. As follows from Remark 24.3, (RC_{PZ}^{S+T}) implies the condition

$$(\widetilde{RC}^{S+T})\ \{(x^* + y^*, x, y, r) : h_S^*(x^*, x) + h_T^*(y^*, y) \leq r\} \text{ is closed}$$

and this further implies (RC^{S+T}). The maximal monotone operators in the example below have as sum a maximal monotone operator, the regularity conditions $(\widetilde{RC}^{S+T})$ and (RC^{S+T}) are fulfilled, but (RC_{PZ}^{S+T}) and, consequently, the generalized interior point conditions from above fail.

Example 25.6. Take $X = \mathbb{R}^2$, equipped with the Euclidean norm $\|\cdot\|_2$, $f, g : \mathbb{R}^2 \to \overline{\mathbb{R}}$, $f = \|\cdot\|_2 + \delta_{\mathbb{R}^2_+}$, $g = \delta_{-\mathbb{R}^2_+}$ and $S = \partial f$, $T = \partial g$. Both S and T are maximal monotone operators and for all $x \in \mathbb{R}^2$ one has $(S + T)(x) = \partial f(x) + \partial g(x) = \mathbb{R}^2$ if $x = 0$, being the empty set, otherwise. This means that $S + T$ is maximal monotone, too. Since f is proper, sublinear and lower semicontinuous and g is the indicator function of a nonempty, convex and closed set, the operators S and T have both unique representative functions, namely

$$h_S(x, x^*) = f(x) + f^*(x^*)\ \forall (x, x^*) \in \mathbb{R}^2 \times \mathbb{R}^2$$

and

$$h_T(x,x^*) = g(x) + g^*(x^*) \ \forall (x,x^*) \in \mathbb{R}^2 \times \mathbb{R}^2,$$

respectively. One can see that $g^* = \delta_{\mathbb{R}^2_+}$ and $f^* = \delta_{\{u\in\mathbb{R}^2:\|u\|_2\leq 1\}-\mathbb{R}^2_+}$. For the set involved in the formulation of the closedness-type regularity condition, we have that

$$\{(x^*+y^*,x,y,r) : f(x)+f^*(x^*)+g(y)+g^*(y^*)\leq r\}$$

$$= \mathbb{R}^2 \times \{(x,y,r) : x\in\mathbb{R}^2_+, y\in -\mathbb{R}^2_+, \|x\|_2 \leq r\}.$$

As this set is closed, both $(\widetilde{RC}^{S+T})$ and (RC^{S+T}) are fulfilled.

Since $\Pr_X(\operatorname{dom} h_S) = \mathbb{R}^2_+$ and $\Pr_X(\operatorname{dom} h_T) = -\mathbb{R}^2_+$, the condition (RC^{S+T}_{PZ}) becomes: $\mathbb{R}^2_+$ is a closed linear subspace in $\mathbb{R}^2$, which shows that this condition fails in this case.

We close this section by making some considerations on the closedness-type regularity conditions treated above. As one can notice from the particular case treated in Example 25.6, even if the generalized interior point conditions fail, $(\widetilde{RC}^{S+T})$ and (RC^{S+T}) are both fulfilled. We have a similar situation in Example 25.3 with the regularity conditions stated for A^*TA. This gives rise to the following open problem.

Open problem 25.7 Do there exist maximal monotone operators $S,T : X \rightrightarrows X^*$ with corresponding representative functions h_S and h_T with $\Pr_X(\operatorname{dom} h_S) \cap \Pr_X(\operatorname{dom} h_T) \neq \emptyset$ such that (RC^{S+T}) is fulfilled, while $(\widetilde{RC}^{S+T})$ fails?

Further, one should also notice that (RC^{S+T}) (but also $(\widetilde{RC}^{S+T})$) provides a family of sufficient conditions for the maximality of $S+T$ that can be obtained by considering different representative functions. Let us denote by $(RC^{S+T}(h_S,h_T))$ the regularity condition we formulated by taking as representative functions h_S and h_T for S and T, respectively. It seems to be challenging to find out what kind of relations do exist between the regularity conditions $(RC^{S+T}(h_S,h_T))$ for different choices of h_S and h_T. Employing in the formulation of $(RC^{S+T}(h_S,h_T))$ the *Fitzpatrick* and the *Penot function*, one obtains the following conditions

$(RC^{S+T}(\varphi_S,\varphi_T))$ $\{(x^*+y^*,x,y,r) : \varphi_S^*(x^*,x)+\varphi_T^*(y^*,y)\leq r\}$ is closed regarding the set $X^*\times\Delta_{X^2}\times\mathbb{R}$,

$(RC^{S+T}(\varphi_S,\varphi_T^*))$ $\{(x^*+y^*,x,y,r) : \varphi_S^*(x^*,x)+\varphi_T(y,y^*)\leq r\}$ is closed regarding the set $X^*\times\Delta_{X^2}\times\mathbb{R}$,

$(RC^{S+T}(\varphi_S^*,\varphi_T))$ $\{(x^*+y^*,x,y,r) : \varphi_S(x,x^*)+\varphi_T^*(y^*,y)\leq r\}$ is closed regarding the set $X^*\times\Delta_{X^2}\times\mathbb{R}$,

and

$(RC^{S+T}(\varphi_S^*,\varphi_T^*))$ $\{(x^*+y^*,x,y,r) : \varphi_S(x,x^*)+\varphi_T(y,y^*)\leq r\}$ is closed regarding the set $X^*\times\Delta_{X^2}\times\mathbb{R}$, each of them guaranteeing that $S+T$ is maximal monotone. The following question appears to be natural.

Open problem 25.8 Is there any relation between the four regularity conditions listed above?

If the answer is negative, it would be important to provide some relevant examples illustrating this fact. As one can see, the generalized interior point regularity condition (RC_{PZ}^{S+T}) is written also in terms of the representative functions of the operators involved. Nevertheless, it makes no sense to raise for it a similar question, as in the case of the closedness-type condition, since for every pair (h_S, h_T) of representative functions, (RC_{PZ}^{S+T}) is equivalent to $0 \in \text{sqri}(D(S) - D(T))$ and this is written in terms of the domains of the operators S and T.

26 Enlargements of Monotone Operators

The notion of *enlargement* of a monotone operator has been introduced in analogy to the notion of ε-subdifferential in the convex analysis. Having $f : X \to \overline{\mathbb{R}}$ a proper, convex and lower semicontinuous function defined on a nonzero Banach space X, assumptions which guarantee that ∂f is a maximal monotone operator, for $\varepsilon \geq 0$ the ε-subdifferential $\partial_\varepsilon f$ is an enlargement of ∂f in the sense that $\partial f(x) \subseteq \partial_\varepsilon f(x)$ for all $x \in X$.

For an arbitrary monotone operator $S : X \rightrightarrows X^*$, the following enlargement has been introduced in [45]: given $\varepsilon \geq 0$, let $S^\varepsilon : X \rightrightarrows X^*$ be for all $x \in X$ defined by

$$S^\varepsilon(x) := \{x^* \in X^* : \langle y^* - x^*, y - x \rangle \geq -\varepsilon \ \forall (y, y^*) \in G(S)\}.$$

This enlargement turned out to have useful applications and properties similar to those of the ε-subdifferential, like local boundedness, demiclosed graph, Lipschitz continuity and the Brøndsted–Rockafellar property (see [50, 51, 72]).

Due to the monotonicity of S, we have $S(x) \subseteq S^\varepsilon(x)$ for all $x \in X$ and $\varepsilon \geq 0$. The operator S^0 needs not be monotone. It is worth noting that $G(S^0)$ is exactly the set of monotonically related elements to $G(S)$, hence S is maximal monotone if and only if $S = S^0$ (see [45, Proposition 2] and [111, Proposition 3.1]). The enlargement S^ε can be characterized via the Fitzpatrick function associated to S:

$$x^* \in S^\varepsilon(x) \Leftrightarrow \varphi_S(x, x^*) \leq \varepsilon + \langle x^*, x \rangle.$$

Motivated by this characterization, the following enlargement for the monotone operator S can be considered (cf. [51, 53]): for a representative function h_S of S we define for all $x \in X$ and $\varepsilon \geq 0$

$$S_{h_S}^\varepsilon(x) := \{x^* \in X^* : h_S(x, x^*) \leq \varepsilon + \langle x^*, x \rangle\}.$$

Obviously, $S^{\varepsilon}_{\varphi_S} = S^{\varepsilon}$ and $G(S) \subseteq G(S^{\varepsilon}_{h_S})$ for all $\varepsilon \geq 0$. Moreover, $S^{\varepsilon}_{h_S}$ has convex strong-closed values and it holds $S^{\varepsilon_1}_{h_S}(x) \subseteq S^{\varepsilon_2}_{h_S}(x)$, provided that $0 \leq \varepsilon_1 \leq \varepsilon_2$. Further, if S is maximal monotone, then, in view of Proposition 23.3, we have for all $x \in X$

$$S(x) \subseteq S^{\varepsilon}_{\psi_S}(x) \subseteq S^{\varepsilon}_{h_S}(x) \subseteq S^{\varepsilon}_{\varphi_S}(x) = S^{\varepsilon}(x)$$

and

$$S(x) \subseteq S^{\varepsilon}_{\psi_S}(x) \subseteq S^{\varepsilon}_{h^*_S}(x) \subseteq S^{\varepsilon}_{\varphi_S}(x) = S^{\varepsilon}(x),$$

where $S^{\varepsilon}_{h^*_S}(x) = \{x^* \in X^* : h^*_S(x^*, x) \leq \varepsilon + \langle x^*, x \rangle\}$, as well as $S = S^0_{\psi_S} = S^0_{h_S} = S^0_{h^*_S} = S^0_{\varphi_S} = S^0$. Let us notice that in case S is a monotone operator and $S = S^0_{h_S}$, where $h_S \neq \varphi_S$, we do not necessarily have that S is maximal monotone.

If $S = \partial f$, where $f : X \to \overline{\mathbb{R}}$ is proper, convex and lower semicontinuous, then for all $x \in X$

$$\partial f(x) \subseteq \partial_{\varepsilon} f(x) \subseteq \partial^{\varepsilon} f(x) := (\partial f)^{\varepsilon}(x)$$

and the inclusions can be strict (see [45, 97]). Moreover, by taking as representative function for ∂f $h : X \times X^* \to \overline{\mathbb{R}}$, $h(x, x^*) = f(x) + f^*(x^*)$, we see that for all $x \in X$ and $\varepsilon \geq 0$ $(\partial f)^{\varepsilon}_h(x) = \partial_{\varepsilon} f(x)$.

As follows from Theorems 7.7 and 7.9, when X is a separated locally convex space and $f, g : X \to \overline{\mathbb{R}}$ are proper, convex and lower semicontinuous functions such that $\operatorname{dom} f \cap \operatorname{dom} g \neq \emptyset$, a necessary and sufficient condition for having

$$\partial_{\varepsilon}(f+g)(x) = \bigcup_{\substack{\varepsilon_1, \varepsilon_2 \geq 0 \\ \varepsilon_1 + \varepsilon_2 = \varepsilon}} \big(\partial_{\varepsilon_1} f(x) + \partial_{\varepsilon_2} g(x)\big) \ \forall x \in X \ \varepsilon \geq 0 \tag{26.1}$$

is that $\operatorname{epi} f^* + \operatorname{epi} g^*$ is closed in $(X^*, \omega(X^*, X)) \times \mathbb{R}$. This equivalence motivates the investigations we make in this section, namely we give a necessary and sufficient condition that ensures a relation similar to (26.1) for the enlargements defined above via an arbitrary representative function.

For the beginning, we prove some preliminary results and start with a theorem which leads to the so-called *bivariate infimal convolution formula*.

Theorem 26.1. *Let X and Y be separated locally convex spaces, $h_1, h_2 : X \times Y \to \overline{\mathbb{R}}$ be proper, convex and lower semicontinuous functions such that $\Pr_X(\operatorname{dom} h_1) \cap \Pr_X(\operatorname{dom} h_2) \neq \emptyset$ and V be a nonempty subset of Y^*. Consider the functions $h_1 \Box_2 h_2 : X \times Y \to \overline{\mathbb{R}}$, $h_1 \Box_2 h_2(x, y) = \inf\{h_1(x, u) + h_2(x, v) : u, v \in Y, u + v = y\}$ and $h^*_1 \Box_1 h^*_2 : X^* \times Y^* \to \overline{\mathbb{R}}$, $(h^*_1 \Box_1 h^*_2)(x^*, y^*) = \inf\{h^*_1(u^*, y^*) + h^*_2(v^*, y^*) : u^*, v^* \in X^*, u^* + v^* = x^*\}$. Then the following conditions are equivalent:*

(i) $(h_1 \Box_2 h_2)^*(x^*, y^*) = (h^*_1 \Box_1 h^*_2)(x^*, y^*)$ and $h^*_1 \Box_1 h^*_2$ is exact at (x^*, y^*) (that is, the infimum in the definition of $(h^*_1 \Box_1 h^*_2)(x^*, y^*)$ is attained) for all $(x^*, y^*) \in X^* \times V$;

(ii) $\{(a^* + b^*, u^*, v^*, r) : h^*_1(a^*, u^*) + h^*_2(b^*, v^*) \leq r\}$ is closed regarding the set $X^* \times \Delta_{V^2} \times \mathbb{R}$ in $(X^*, \omega(X^*, X)) \times (Y^*, \omega(Y^*, Y)) \times (Y^*, \omega(Y^*, Y)) \times \mathbb{R}$.

Proof. Take an arbitrary $(x^*, y^*) \in X^* \times Y^*$. The following equality can be easily derived:

$$(h_1 \Box_2 h_2)^*(x^*, y^*) = \sup_{x \in X, u, v \in Y} \{\langle x^*, x\rangle + \langle y^*, u + v\rangle - h_1(x, u) - h_2(x, v)\}. \tag{26.2}$$

Define now the functions $F, G : X \times Y \times Y \to \overline{\mathbb{R}}$, $F(x, u, v) = h_1(x, u)$ and $G(x, u, v) = h_2(x, v)$, $(x, u, v) \in X \times Y \times Y$. It holds $(h_1 \Box_2 h_2)^*(x^*, y^*) = (F + G)^*(x^*, y^*, y^*)$. For all $(x^*, u^*, v^*) \in X^* \times Y^* \times Y^*$ the conjugate functions $F^*, G^* : X^* \times Y^* \times Y^* \to \overline{\mathbb{R}}$ look like

$$F^*(x^*, u^*, v^*) = \begin{cases} h_1^*(x^*, u^*), & \text{if } v^* = 0, \\ +\infty, & \text{otherwise} \end{cases}$$

and

$$G^*(x^*, u^*, v^*) = \begin{cases} h_2^*(x^*, v^*), & \text{if } u^* = 0, \\ +\infty, & \text{otherwise,} \end{cases}$$

respectively. Further, we have $(F^* \Box G^*)(x^*, y^*, y^*) = (h_1^* \Box_1 h_2^*)(x^*, y^*)$. Hence the condition (i) is fulfilled if and only if $(F + G)^*(x^*, y^*, y^*) = (F^* \Box G^*)(x^*, y^*, y^*)$ and $(F^* \Box G^*)$ is exact at (x^*, y^*, y^*) for all $(x^*, y^*, y^*) \in X^* \times \Delta_{V^2}$. In view of Theorem 9.2, this is further equivalent to epi F^* + epi G^* is closed regarding the set $X^* \times \Delta_{V^2} \times \mathbb{R}$ in $(X^*, \omega(X^*, X)) \times (Y^*, \omega(Y^*, Y)) \times (Y^*, \omega(Y^*, Y)) \times \mathbb{R}$. Finally, the equality epi F^* + epi $G^* = \{(a^* + b^*, u^*, v^*, r) : h_1^*(a^*, u^*) + h_2^*(b^*, v^*) \leq r\}$, the proof of which presents no difficulty, gives the desired result. □

For the particular case when $V := Y^*$, we obtain the following necessary and sufficient condition for the bivariate infimal convolution formula, i.e. the formula encountered in statement (i) below.

Corollary 26.2. *Let X and Y be separated locally convex spaces, $h_1, h_2 : X \times Y \to \overline{\mathbb{R}}$ be proper, convex and lower semicontinuous functions such that* $\Pr_X(\operatorname{dom} h_1) \cap \Pr_X(\operatorname{dom} h_2) \neq \emptyset$. *The following statements are equivalent:*

(i) $(h_1 \Box_2 h_2)^* = h_1^* \Box_1 h_2^*$ *and* $h_1^* \Box_1 h_2^*$ *is exact;*

(ii) $\{(a^* + b^*, u^*, v^*, r) : h_1^*(a^*, u^*) + h_2^*(b^*, v^*) \leq r\}$ *is closed regarding the set* $X^* \times \Delta_{Y^{*2}} \times \mathbb{R}$ *in* $(X^*, \omega(X^*, X)) \times (Y^*, \omega(Y^*, Y)) \times (Y^*, \omega(Y^*, Y)) \times \mathbb{R}$.

Remark 26.3. A generalized interior point condition, which ensures relation (i) in Corollary 26.2, was given in [121, Theorem 4.2], namely

$$0 \in \operatorname{sqri}(\Pr_X(\operatorname{dom} h_1) - \Pr_X(\operatorname{dom} h_2)).$$

Unlike (ii), which is a necessary and sufficient condition for (i), the condition above is only sufficient. To see this, it is enough to take the functions h_S and h_T from Example 25.6. As shown there,

$$0 \notin \operatorname{sqri}(\Pr_X(\operatorname{dom} h_S) - \Pr_X(\operatorname{dom} h_T)),$$

but

$$\{(a^* + b^*, u^*, v^*, r) : h_S^*(a^*, u^*) + h_T^*(b^*, v^*) \le r\}$$
$$= \mathbb{R}^2 \times \{(u^*, v^*, r) : u^* \in \mathbb{R}^2_+, v^* \in -\mathbb{R}^2_+, \|u^*\|_2 \le r\},$$

which is a closed set.

By taking in Theorem 26.1 $Y := X^*$ and $V := X$, where X is supposed to be a normed space, so that $V = X$ can be seen as a subspace of $Y^* = X^{**}$, we obtain the following result.

Corollary 26.4. *Let X be a normed space and $h_1, h_2 : X \times X^* \to \overline{\mathbb{R}}$ be proper, convex and lower semicontinuous functions such that* $\Pr_X(\operatorname{dom} h_1) \cap \Pr_X(\operatorname{dom} h_2) \neq \emptyset$. *The following statements are equivalent:*

(i) $(h_1 \Box_2 h_2)^*(x^*, x) = (h_1^* \Box_1 h_2^*)(x^*, x)$ *and* $h_1^* \Box_1 h_2^*$ *is exact at* (x^*, x) *for all* $(x^*, x) \in X^* \times X$;

(ii) $\{(a^* + b^*, u^{**}, v^{**}, r) : h_1^*(a^*, u^{**}) + h_2^*(b^*, v^{**}) \le r\}$ *is closed regarding the set* $X^* \times \Delta_{X^2} \times \mathbb{R}$ *in* $(X^*, \omega(X^*, X)) \times (X^{**}, \omega(X^{**}, X^*)) \times (X^{**}, \omega(X^{**}, X^*)) \times \mathbb{R}$.

We return to the setting considered at the beginning of the section, namely with X a nonzero real Banach space, and prove that condition (ii) in Corollary 26.4 is sufficient for having that $(h_S \Box_2 h_T)^*$ is representative for $S + T$ in case h_S and h_T are representative functions for S and T, respectively. Moreover, Theorem 26.5 offers an alternative proof for the statement in Theorem 25.4.

Theorem 26.5. *Let $S, T : X \rightrightarrows X^*$ be maximal monotone operators with representative functions h_S and h_T, respectively, such that* $\Pr_X(\operatorname{dom} h_S) \cap \Pr_X(\operatorname{dom} h_T) \neq \emptyset$ *and consider the function $h : X \times X^* \to \overline{\mathbb{R}}$, $h = (h_S \Box_2 h_T)^{*\top}$. If*

$\{(a^* + b^*, u^{**}, v^{**}, r) : h_S^*(a^*, u^{**}) + h_T^*(b^*, v^{**}) \le r\}$ *is closed regarding the set* $X^* \times \Delta_{X^2} \times \mathbb{R}$ *in* $(X^*, \omega(X^*, X)) \times (X^{**}, \omega(X^{**}, X^*)) \times (X^{**}, \omega(X^{**}, X^*)) \times \mathbb{R}$,

then h is a representative function of the monotone operator $S + T$. If, additionally, X is reflexive, then $S + T$ is a maximal monotone operator.

Proof. The function h is obviously convex and strong-weak* lower semicontinuous, hence strong lower semicontinuous. By applying Corollary 26.4, we obtain $h(x, x^*) = (h_S^* \Box_1 h_T^*)(x^*, x)$ and $h_S^* \Box_1 h_T^*$ is exact at (x^*, x) for all $(x^*, x) \in X^* \times X$. By using Proposition 23.3 we have for all $(x, x^*) \in X \times X^*$ that $h(x, x^*) = h_S^* \Box_1 h_T^*(x^*, x) = \inf\{h_S^*(u^*, x) + h_T^*(v^*, x) : u^*, v^* \in X^*, u^* + v^* = x^*\} \ge \inf\{\langle u^*, x\rangle + \langle v^*, x\rangle : u^*, v^* \in X^*, u^* + v^* = x^*\} = \langle x^*, x\rangle$, hence $h \ge c$.

It remains to show that $G(S + T) \subseteq \{(x, x^*) \in X \times X^* : h(x, x^*) = \langle x^*, x\rangle\}$. Take an arbitrary $(x, x^*) \in G(S + T)$. Then there exist $u^* \in S(x)$ and $v^* \in T(x)$ such that $x^* = u^* + v^*$. By employing once more Proposition 23.3, we obtain

$$\langle x^*, x\rangle \le h(x, x^*) = h_S^* \Box_1 h_T^*(x^*, x)$$

$$\leq h_S^*(u^*, x) + h_T^*(v^*, x) = \langle u^*, x\rangle + \langle v^*, x\rangle = \langle x^*, x\rangle,$$

thus the inclusion is proved.

Actually, we have more, namely that

$$G(S+T) = \{(x, x^*) \in X \times X^* : h(x, x^*) = \langle x^*, x\rangle\}. \tag{26.3}$$

Take an arbitrary $(x, x^*) \in X \times X^*$ such that $h(x, x^*) = \langle x^*, x\rangle$. By Corollary 26.4, there exist $u^*, v^* \in X^*$, $u^* + v^* = x^*$, such that

$$h_S^*(u^*, x) + h_T^*(v^*, x) = \langle u^*, x\rangle + \langle v^*, x\rangle. \tag{26.4}$$

The functions h_S and h_T being representative, by Proposition 23.3, we have $h_S^*(u^*, x) \geq \langle u^*, x\rangle$ and $h_T^*(v^*, x) \geq \langle v^*, x\rangle$, hence, in view of (26.4), the inequalities above must be fulfilled as equalities. By applying again Proposition 23.3, we get $u^* \in S(x)$ and $v^* \in T(x)$, so $x^* = u^* + v^* \in S(x) + T(x) = (S+T)(x)$ and (26.3) is fulfilled.

Suppose now that X is a reflexive Banach space. Since the weak closure of a convex set is exactly the strong closure of the same set, the regularity condition becomes the condition (RC^{S+T}) given in the previous section. We give in the following a different proof of the fact that $S + T$ is maximal monotone.

As h_S and h_T are representative functions we have $h_S \square_2 h_T \geq c$. The duality product is continuous in the strong topology on $X \times X^*$, so it follows $\text{cl}_{\|\cdot\| \times \|\cdot\|_*}(h_S \square_2 h_T) \geq c$. Therefore, $h^{*\top} = \text{cl}_{\|\cdot\| \times \|\cdot\|_*}(h_S \square_2 h_T) \geq c$. The conclusion follows now by combining Theorem 23.5 with relation (26.3). □

We state now the result for the enlargements of $S + T$ we announced at the beginning of the section (cf. [15]).

Theorem 26.6. *Let $S, T : X \rightrightarrows X^*$ be maximal monotone operators with representative functions h_S and h_T, respectively, such that $\Pr_X(\text{dom}\, h_S) \cap \Pr_X(\text{dom}\, h_T) \neq \emptyset$ and consider again $h : X \times X^* \to \overline{\mathbb{R}}$, $h = (h_S \square_2 h_T)^{*\top}$. Then the following statements are equivalent:*

(i) $\{(a^* + b^*, u^{**}, v^{**}, r) : h_S^*(a^*, u^{**}) + h_T^*(b^*, v^{**}) \leq r\}$ is closed regarding the set $X^* \times \Delta_{X^2} \times \mathbb{R}$ in $(X^*, \omega(X^*, X)) \times (X^{**}, \omega(X^{**}, X^*)) \times (X^{**}, \omega(X^{**}, X^*)) \times \mathbb{R}$;

(ii) $(S+T)_h^\varepsilon(x) = \bigcup\limits_{\substack{\varepsilon_1, \varepsilon_2 \geq 0 \\ \varepsilon_1 + \varepsilon_2 = \varepsilon}} \left(S_{h_S^*}^{\varepsilon_1}(x) + T_{h_T^*}^{\varepsilon_2}(x)\right)$ for all $\varepsilon \geq 0$ and $x \in X$,

where $(S+T)_h^\varepsilon(x) := \{x^* \in X^* : h(x, x^*) \leq \varepsilon + \langle x^*, x\rangle\}$ for every $x \in X$ and $\varepsilon \geq 0$.

Proof. Before proving the equivalence, we want to notice that, in view of Theorem 26.5, when (i) is fulfilled, then h is a representative function of the operator $S + T$, hence the notation $(S+T)_h^\varepsilon(x) := \{x^* \in X^* : h(x, x^*) \leq \varepsilon + \langle x^*, x\rangle\}$

is justified. As we show in the following, when condition (ii) is true, then (i) is also fulfilled, thus also in this case the use of this notation makes sense.

Suppose now that (i) is fulfilled and take $x \in X$ and $\varepsilon \geq 0$. We show first the inclusion

$$\bigcup_{\substack{\varepsilon_1,\varepsilon_2\geq 0\\ \varepsilon_1+\varepsilon_2=\varepsilon}} \left(S^{\varepsilon_1}_{h^*_S}(x) + T^{\varepsilon_2}_{h^*_T}(x)\right) \subseteq (S+T)^{\varepsilon}_h(x). \tag{26.5}$$

Indeed, take $\varepsilon_1, \varepsilon_2 \geq 0, \varepsilon_1+\varepsilon_2 = \varepsilon, u^* \in S^{\varepsilon_1}_{h^*_S}(x)$ and $v^* \in T^{\varepsilon_2}_{h^*_T}(x)$. Then $h(x, u^*+v^*) = (h_S\Box_2 h_T)^*(u^*+v^*, x) \leq h^*_S\Box_1 h^*_T(u^*+v^*, x) \leq h^*_S(u^*, x)+h^*_T(v^*, x) \leq \varepsilon_1 + \langle u^*, x\rangle + \varepsilon_2 + \langle v^*, x\rangle = \varepsilon + \langle u^*+v^*, x\rangle$, hence $u^*+v^* \in (S+T)^{\varepsilon}_h(x)$, that is, the inclusion (26.5) is true. Let us mention that this inclusion is always fulfilled, as there is no need of (i) to prove it.

However, for the opposite inclusion we use condition (i). Take $x^* \in (S+T)^{\varepsilon}_h(x)$ or, equivalently, $(h_S\Box_2 h_T)^*(x^*, x) \leq \varepsilon + \langle x^*, x\rangle$. By Corollary 26.4, we get $h^*_S\Box_1 h^*_T(x^*, x) \leq \varepsilon + \langle x^*, x\rangle$ and the infimum in the definition of $h^*_S\Box_1 h^*_T(x^*, x)$ is attained. Hence there exist $u^*, v^* \in X^*$ such that $u^*+v^* = x^*$ and

$$h^*_S(u^*, x) + h^*_T(v^*, x) \leq \varepsilon + \langle u^*, x\rangle + \langle v^*, x\rangle. \tag{26.6}$$

Take $\varepsilon_1 := h^*_S(u^*, x) - \langle u^*, x\rangle$ and $\varepsilon_2 := \varepsilon - \varepsilon_1$. By (26.6) and Proposition 23.3 we get $\varepsilon_1 \geq 0$ and $\varepsilon_2 \geq h^*_T(v^*, x) - \langle v^*, x\rangle \geq 0$. Thus $u^* \in S^{\varepsilon_1}_{h^*_S}(x)$ and $v^* \in T^{\varepsilon_2}_{h^*_T}(x)$, that is

$$x^* = u^* + v^* \in \bigcup_{\substack{\varepsilon_1,\varepsilon_2\geq 0\\ \varepsilon_1+\varepsilon_2=\varepsilon}} \left(S^{\varepsilon_1}_{h^*_S}(x) + T^{\varepsilon_2}_{h^*_T}(x)\right),$$

so (ii) is fulfilled.

Conversely, assume that (ii) is true. We prove first

$$h(x, x^*) \geq \langle x^*, x\rangle \ \forall (x, x^*) \in X \times X^*. \tag{26.7}$$

We suppose that there exists $(x_0, x^*_0) \in X \times X^*$ such that $h(x_0, x^*_0) \leq \langle x^*_0, x_0\rangle$. By using (ii) for $\varepsilon := 0$ we obtain $x^*_0 \in (S+T)^0_h(x_0) = S^0_{h^*_S}(x_0)+T^0_{h^*_T}(x_0) = S(x_0)+T(x_0)$. Hence there exist $u^*_0 \in S(x_0)$ and $v^*_0 \in T(x_0)$ such that $x^*_0 = u^*_0+v^*_0$. From Proposition 23.3, we obtain $h_S(x_0, u^*_0) = \langle u^*_0, x_0\rangle$ and $h_T(x_0, v^*_0) = \langle v^*_0, x_0\rangle$ and further

$$\begin{aligned} h(x_0, x^*_0) &= \sup_{x\in X, u^*, v^*\in X^*} \{\langle x^*_0, x\rangle + \langle u^*, x_0\rangle + \langle v^*, x_0\rangle - h_S(x, u^*) - h_T(x, v^*)\} \\ &\geq \langle x^*_0, x_0\rangle + \langle u^*_0, x_0\rangle + \langle v^*_0, x_0\rangle - h_S(x_0, u^*_0) - h_T(x_0, v^*_0) = \langle x^*_0, x_0\rangle, \end{aligned}$$

thus (26.7) is fulfilled.

In view of Corollary 26.4, it is sufficient to show that $h(x, x^*) = h^*_S\Box_1 h^*_T(x^*, x)$ and $h^*_S\Box_1 h^*_T$ is exact at (x^*, x) for all $(x^*, x) \in X^*\times X$. Take an arbitrary $(x^*, x) \in X^* \times X$. The inequality

$$h(x,x^*) \le (h_S^* \square_1 h_T^*)(x^*,x) \tag{26.8}$$

is always true. When $h(x,x^*) = +\infty$, there is nothing to be proved. The condition $\Pr_X(\operatorname{dom} h_S) \cap \Pr_X(\operatorname{dom} h_T) \neq \emptyset$ ensures that $h(x,x^*) > -\infty$, so we may suppose that $h(x,x^*) \in \mathbb{R}$. Let us denote by $r := h(x,x^*)$. We have $h(x,x^*) = \langle x^*,x\rangle + (r - \langle x^*,x\rangle)$. With $\varepsilon := r - \langle x^*,x\rangle \ge 0$ (cf. (26.7)) we obtain that $x^* \in (S+T)_h^\varepsilon(x)$. Since (ii) is true, there exist $\varepsilon_1,\varepsilon_2 \ge 0$, $\varepsilon_1+\varepsilon_2 = \varepsilon$ and $u^* \in S_{h_S^*}^{\varepsilon_1}(x)$ and $v^* \in T_{h_T^*}^{\varepsilon_2}(x)$, respectively, such that $x^* = u^* + v^*$. Further, adding the inequalities

$$h_S^*(u^*,x) \le \varepsilon_1 + \langle u^*,x\rangle$$

and

$$h_T^*(v^*,x) \le \varepsilon_2 + \langle v^*,x\rangle$$

we obtain

$$h_S^*(u^*,x) + h_T^*(v^*,x) \le \varepsilon_1 + \varepsilon_2 + \langle u^*+v^*,x\rangle = r = h(x,x^*),$$

hence, in view of (26.8) we get $h(x,x^*) = h_S^*(u^*,x) + h_T^*(v^*,x) = h_S^* \square_1 h_2^*(x^*,x)$ and the proof is complete. □

Remark 26.7. Due to Remark 26.3 and 26.6, one has that all the generalized interior point regularity conditions provided in Remark 25.5 for the maximal monotonicity of $S+T$ are sufficient conditions for having

$$(S+T)_h^\varepsilon(x) = \bigcup_{\substack{\varepsilon_1,\varepsilon_2\ge 0\\ \varepsilon_1+\varepsilon_2=\varepsilon}} \left(S_{h_S^*}^{\varepsilon_1}(x) + T_{h_T^*}^{\varepsilon_2}(x)\right) \ \forall \varepsilon \ge 0\ \forall x \in X.$$

That these conditions are only sufficient for this relation, can be seen by considering the operators S and T and the representative functions h_S and h_T from Example 25.6.

In the following, we show that the equivalence proved in Theorem 7.9 in separated locally convex spaces follows when X is a Banach space as a particular instance of Theorem 26.6.

Corollary 26.8. *Let $f,g : X \to \overline{\mathbb{R}}$ be proper, convex and lower semicontinuous functions such that* $\operatorname{dom} f \cap \operatorname{dom} g \neq \emptyset$. *The following statements are equivalent:*

(i) $\operatorname{epi} f^* + \operatorname{epi} g^*$ *is closed in* $(X^*,\omega(X^*,X)) \times \mathbb{R}$;
(ii) $\partial_\varepsilon(f+g)(x) = \bigcup_{\substack{\varepsilon_1,\varepsilon_2\ge 0\\ \varepsilon_1+\varepsilon_2=\varepsilon}} \left(\partial_{\varepsilon_1} f(x) + \partial_{\varepsilon_2} g(x)\right)$ *for all* $\varepsilon \ge 0$ *and* $x \in X$.

Proof. Consider the functions $h_1,h_2 : X \times X^* \to \overline{\mathbb{R}}$, $h_1(x,x^*) = f(x) + f^*(x^*)$ and $h_2(x,x^*) = g(x) + g^*(x^*)$ for all $(x,x^*) \in X \times X^*$. We have $h_1^*(x^*,x^{**}) = f^{**}(x^{**}) + f^*(x^*)$ and $h_2^*(x^*,x^{**}) = g^{**}(x^{**}) + g^*(x^*)$ for all $(x^*,x^{**}) \in$

$X^* \times X^{**}$. Further, $(h_1 \Box_2 h_2)^*(x^*, x) = (h_1^* \Box_1 h_2^*)(x^*, x)$ and $h_1^* \Box_1 h_2^*$ is exact at (x^*, x) for all $(x^*, x) \in X^* \times X$ is fulfilled if and only if $(f + g)^* = f^* \Box g^*$ and $f^* \Box g^*$ is exact. Therefore, by Theorem 7.7, the condition $\operatorname{epi} f^* + \operatorname{epi} g^*$ is closed in $(X^*, \omega(X^*, X)) \times \mathbb{R}$ is nothing else than condition (i) in Corollary 26.4. This is further equivalent to

$\{(a^* + b^*, u^{**}, v^{**}, r) : h_1^*(a^*, u^{**}) + h_2^*(b^*, v^{**}) \leq r\}$ is closed regarding the set $X^* \times \Delta_{X^2} \times \mathbb{R} \text{in} (X^*, \omega(X^*, X)) \times (X^{**}, \omega(X^{**}, X^*)) \times (X^{**}, \omega(X^{**}, X^*)) \times \mathbb{R}$.

Since h_1 and h_2 are representative functions of the maximal monotone operators ∂f and ∂g, respectively, we obtain, by Theorem 26.6, that (i) is fulfilled if and only if for all $\varepsilon \geq 0$ and $x \in X$ it holds

$$(\partial f + \partial g)_h^\varepsilon(x) = \bigcup_{\substack{\varepsilon_1, \varepsilon_2 \geq 0 \\ \varepsilon_1 + \varepsilon_2 = \varepsilon}} \left((\partial f)_{h_1^*}^{\varepsilon_1}(x) + (\partial g)_{h_2^*}^{\varepsilon_2}(x) \right),$$

where $h : X \times X^* \to \overline{\mathbb{R}}$, $h(x, x^*) = (h_1 \Box_2 h_2)^*(x^*, x) = (f + g)(x) + (f + g)^*(x^*)$. Taking into consideration that $(\partial f + \partial g)_h^\varepsilon(x) = \{x^* \in X^* : (f + g)(x) + (f + g)^*(x^*) \leq \varepsilon + \langle x^*, x \rangle\} = \partial_\varepsilon (f + g)(x)$ and $(\partial f)_{h_1^*}^{\varepsilon_1}(x) = \partial_{\varepsilon_1} f(x)$, respectively, $(\partial g)_{h_2^*}^{\varepsilon_2}(x) = \partial_{\varepsilon_2} g(x)$, we get the desired result.

□

Remark 26.9. In reflexive Banach spaces one can deduce the equivalence in Corollary 26.8 also from [53, Theorem 6.9]. On the other hand, following the approach presented in this section, one can give a similar result to Theorem 26.6 by considering instead of the sum of two maximal monotone operators the operator $S + A^*TA$, where X, Y are Banach spaces, $S : X \rightrightarrows X^*$ and $T : Y \rightrightarrows Y^*$ are maximal monotone operators and $A : X \to Y$ is a linear continuous operator.

References

1. Aït Mansour M, Metrane A, Théra M (2006): *Lower semicontinuous regularization for vector-valued mappings*. Journal of Global Optimization 35(2), 283–309
2. Attouch H, Brézis H (1986): *Duality for the sum of convex functions in general Banach spaces*. In: Barroso JA (ed) Aspects of Mathematics and Its Applications. North-Holland, Amsterdam, 125–133
3. Attouch H, Riahi H, Théra M (1996): *Somme ponctuelle d'opérateurs maximaux monotones*. Serdica – Mathematical Journal 22(3), 165–190
4. Bauschke HH (1999): *Proof of a conjecture by Deutsch, Li, and Swetits on duality of optimization problems*. Journal of Optimization Theory and Applications 102(3), 697–703
5. Bauschke HH, Borwein JM, Tseng P (2000): *Bounded linear regularity, strong CHIP, and CHIP are distinct properties*. Journal of Convex Analysis 7(2), 395–412
6. Bauschke HH, McLaren DA, Sendov HS (2006): *Fitzpatrick functions: Inequalities, examples and remarks on a problem by S. Fitzpatrick*. Journal of Convex Analysis 13(3–4), 499–523
7. Borwein JM (2006): *Maximal monotonicity via convex analysis*. Journal of Convex Analysis 13(3–4), 561–586
8. Borwein JM (2007): *Maximality of sums of two maximal monotone operators in general Banach space*. Proceedings of the American Mathematical Society 135(12), 3917–3924
9. Borwein JM, Goebel R (2003): *Notions of relative interior in Banach spaces*. Journal of Mathematical Sciences 115(4), 2542–2553
10. Borwein JM, Jeyakumar V, Lewis A, Wolkowicz H (1988): *Constrained approximation via convex programming*. Unpublished preprint, University of Waterloo
11. Borwein JM, Lewis AS (1992): *Partially finite convex programming, part I: Quasi relative interiors and duality theory*. Mathematical Programming 57(1), 15–48
12. Borwein JM, Lewis AS (2006): *Convex Analysis and Nonlinear Optimization*. Springer, New York
13. Borwein JM, Zhu QJ (2005): *Techniques of Variational Analysis*. Springer, New York
14. Boţ RI (2003): *Duality and Optimality in Multiobjective Optimization*. Ph.D. Thesis, Faculty of Mathematics, Chemnitz University of Technology
15. Boţ RI, Csetnek ER (2008): *An application of the bivariate inf-convolution formula to enlargements of monotone operators*. Set-Valued Analysis 16(7–8), 983–997
16. Boţ RI, Csetnek ER (2009): *On an open problem regarding totally Fenchel unstable functions*. Proceedings of the American Mathematical Society 137(5), 1801–1805
17. Boţ RI, Csetnek ER, Moldovan A (2008): *Revisiting some duality theorems via the quasirelative interior in convex optimization*. Journal of Optimization Theory and Applications 139(1), 67–84
18. Boţ RI, Csetnek ER, Wanka G (2007): *A new condition for maximal monotonicity via representative functions*. Nonlinear Analysis: Theory, Methods & Applications 67(8), 2390–2402
19. Boţ RI, Csetnek ER, Wanka G (2008): *Sequential optimality conditions in convex programming via perturbation approach*. Journal of Convex Analysis 15(1), 149–164
20. Boţ RI, Csetnek ER, Wanka G (2008): *Regularity conditions via quasi-relative interior in convex programming*. SIAM Journal on Optimization 19(1), 217–233

21. Boţ RI, Grad SM (2008): *Regularity conditions for formulae of biconjugate functions.* Taiwanese Journal of Mathematics 12(8), 1921–1942
22. Boţ RI, Grad SM, Wanka G (2006): *Fenchel–Lagrange versus geometric duality in convex optimization.* Journal of Optimization Theory and Applications 129(1), 33–54
23. Boţ RI, Grad SM, Wanka G (2006): *Maximal monotonicity for the precomposition with a linear operator.* SIAM Journal on Optimization 17(4), 1239–1252
24. Boţ RI, Grad SM, Wanka G (2009): *New regularity conditions for Lagrange and Fenchel–Lagrange duality in infinite dimensional spaces.* Mathematical Inequalities & Applications 12(1), 171–189
25. Boţ RI, Grad SM, Wanka G (2007): *A general approach for studying duality in multiobjective optimization.* Mathematical Methods of Operations Research (ZOR) 65(3), 417–444
26. Boţ RI, Grad SM, Wanka G (2007): *New constraint qualification and conjugate duality for composed convex optimization problems.* Journal of Optimization Theory and Applications 135(2), 241–255
27. Boţ RI, Grad SM, Wanka G (2009): *Generalized Moreau–Rockafellar results for composed convex functions.* Optimization 58(7), 917–933
28. Boţ RI, Grad SM, Wanka G (2007): *Weaker constraint qualifications in maximal monotonicity.* Numerical Functional Analysis and Optimization 28(1–2), 27–41
29. Boţ RI, Grad SM, Wanka G (2008): *A new constraint qualification for the formula of the subdifferential of composed convex functions in infinite dimensional spaces.* Mathematische Nachrichten 281(8), 1088–1107
30. Boţ RI, Grad SM, Wanka G (2008): *New regularity conditions for strong and total Fenchel–Lagrange duality in infinite dimensional spaces.* Nonlinear Analysis: Theory, Methods & Applications 69(1), 323–336
31. Boţ RI, Grad SM, Wanka G (2008): *On strong and total Lagrange duality for convex optimization problems.* Journal of Mathematical Analysis and Applications 337(2), 1315–1325
32. Boţ RI, Hodrea IB, Wanka G (2006): *Farkas-type results for inequality systems with composed convex functions via conjugate duality.* Journal of Mathematical Analysis and Applications 322(1), 316–328
33. Boţ RI, Hodrea IB, Wanka G (2008): *ε-Optimality conditions for composed convex optimization problems.* Journal of Approximation Theory 153(1), 108–121
34. Boţ RI, Kassay G, Wanka G (2005): *Strong duality for generalized convex optimization problems.* Journal of Optimization Theory and Applications 127(1), 45–70
35. Boţ RI, Löhne A (2008): *On totally Fenchel unstable functions in finite dimensional spaces.* Mathematical Programming, DOI 10.1007/s10107-009-0311-8
36. Boţ RI, Wanka G (2004): *An analysis of some dual problems in multiobjective optimization (I).* Optimization 53(3), 281–300
37. Boţ RI, Wanka G (2004): *An analysis of some dual problems in multiobjective optimization (II).* Optimization 53(3), 301–324
38. Boţ RI, Wanka G (2005): *Farkas-type results with conjugate functions.* SIAM Journal on Optimization 15(2), 540–554
39. Boţ RI, Wanka G (2006): *An alternative formulation for a new closed cone constraint qualification.* Nonlinear Analysis: Theory, Methods & Applications 64(6), 1367–1381
40. Boţ RI, Wanka G (2006): *A weaker regularity condition for subdifferential calculus and Fenchel duality in infinite dimensional spaces.* Nonlinear Analysis: Theory, Methods & Applications 64(12), 2787–2804
41. Boţ RI, Wanka G (2006): *Farkas-type results for max-functions and applications.* Positivity 10(4), 761–777
42. Boţ RI, Wanka G (2008): *The conjugate of the pointwise maximum of two convex functions revisited.* Journal of Global Optimization 41(4), 625–632
43. Bourgain J (1978): *A geometric characterization of the Radon–Nikodým property in Banach spaces.* Compositio Mathematica 36(1), 3–6
44. Burachik RS, Fitzpatrick SP (2005): *On a family of convex functions associated to subdifferentials.* Journal of Nonlinear and Convex Analysis 6(1), 165–171

45. Burachik RS, Iusem AN, Svaiter BF (1997): *Enlargement of monotone operators with applications to variational inequalities*. Set-Valued Analysis 5(2), 159–180
46. Burachik RS, Jeyakumar V (2005): *A dual condition for the convex subdifferential sum formula with applications*. Journal of Convex Analysis 12(2), 279–290
47. Burachik RS, Jeyakumar V (2005): *A new geometric condition for Fenchel's duality in infinite dimensional spaces*. Mathematical Programming 104(2–3), 229–233
48. Burachik RS, Jeyakumar V (2005): *A simpler closure condition for the normal cone intersection formula*. Proceedings of the American Mathematical Society 133(6), 1741–1748
49. Burachik RS, Jeyakumar V, Wu ZY (2006): *Necessary and sufficient conditions for stable conjugate duality*. Nonlinear Analysis: Theory, Methods & Applications 64(9), 1998–2006
50. Burachik RS, Svaiter BF (1999): *ϵ-Enlargements of maximal monotone operators in Banach spaces*. Set-Valued Analysis 7(2), 117–132
51. Burachik RS, Svaiter BF (2002): *Maximal monotone operators, convex functions and a special family of enlargements*. Set-Valued Analysis 10(4), 297–316
52. Burachik RS, Svaiter BF (2003): *Maximal monotonicity, conjugation and duality product*. Proceedings of the American Mathematical Society 131(8), 2379–2383
53. Burachik RS, Svaiter BF (2006): *Operating enlargements of monotone operators: new connections with convex functions*. Pacific Journal of Optimization 2(3), 425–445
54. Cammaroto F, Di Bella B (2005): *Separation theorem based on the quasirelative interior and application to duality theory*. Journal of Optimization Theory and Applications 125(1), 223–229
55. Chakrabarty AK, Shunmugaraj P, Zălinescu C (2007): *Continuity properties for the subdifferential and ε-subdifferential of a convex function and its conjugate*. Journal of Convex Analysis 14(3), 479–514
56. Chu LJ (1996): *On the sum of monotone operators*. The Michigan Mathematical Journal 43(2), 273–289
57. Combari C, Laghdir M, Thibault L (1994): *Sous-différentiels de fonctions convexes composées*. Annales des Sciences Mathématiques du Québec 18(2), 119–148
58. Combari C, Laghdir M, Thibault L (1996): *A note on subdifferentials of convex composite functionals*. Archiv der Mathematik (Basel) 67(3), 239–252
59. Daniele P, Giuffrè S (2007): *General infinite dimensional duality theory and applications to evolutionary network equilibrium problems*. Optimization Letters 1(3), 227–243
60. Daniele P, Giuffrè S, Idone G, Maugeri A (2007): *Infinite dimensional duality and applications*. Mathematische Annalen 339(1), 221–239
61. Deutsch F, Li W, Swetits J (1999): *Fenchel duality and the strong conical hull intersection property*. Journal of Optimization Theory and Applications 102(3), 681–695
62. Deutsch F, Li W, Ward DJ (1997): *A dual approach to constrained interpolation from a convex subset of Hilbert space*. Journal of Approximation Theory 90(3), 385–414
63. Deutsch F, Li W, Ward DJ (1999): *Best approximation from the intersection of a closed convex set and a polyhedron in Hilbert space, weak Slater conditions, and the strong conical hull intersection property*. SIAM Journal on Optimization 10(1), 252–268
64. Dinh N, Goberna MA, López MA, Son TQ (2007): *New Farkas-type constraint qualifications in convex infinite programming*. ESAIM Control, Optimisation and Calculus of Variations 13(3), 580–597
65. Dinh N, Mordukhovich BS, Nghia TTA (2008): *Subdifferentials of value functions and optimality conditions for some classes of DC and bilevel infinite and semi-infinite programs*. Mathematical Programming, DOI 10.1007/s10107-009-0323-4
66. Dinh N, Vallet G, Nghia TTA (2008): *Farkas-type results and duality for DC programs with convex constraints*. Journal of Convex Analysis 15(2), 235–262
67. Ekeland I, Temam R (1976): *Convex Analysis and Variational Problems*. North-Holland, Amsterdam
68. Ernst E, Théra M (2007): *Boundary half-strips and the strong CHIP*. SIAM Journal on Optimization 18(3), 834–852
69. Fajardo MD, López MA (1999): *Locally Farkas–Minkowski systems in convex semi-infinite programming*. Journal of Optimization Theory and Applications 103(2), 313–335

70. Fitzpatrick SP (1988): *Representing monotone operators by convex functions*. In: Fitzpatrick SP, Giles JR (eds) Workshop/miniconference on Functional Analysis and Optimization: Canberra, 8–24 August, 1988. Proceedings of the Centre for Mathematical Analysis 20, Australian National University, Canberra, 59–65
71. Fitzpatrick SP, Simons S (2001): *The conjugates, compositions and marginals of convex functions*, Journal of Convex Analysis 8(2), 423–446
72. García Y, Lassonde M, Revalski JP (2006): *Extended sums and extended compositions of monotone operators*. Journal of Convex Analysis 13(3–4), 721–738
73. Giannessi F (2005): *Constrained Optimization and Image Space Analysis, Vol. 1. Separation of Sets and Optimality Conditions*. Mathematical Concepts and Methods in Science and Engineering 49. Springer, New York
74. Goberna MA, Jeyakumar V, López MA (2008): *Necessary and sufficient constraint qualifications for solvability of systems of infinite convex inequalities*. Nonlinear Analysis: Theory, Methods & Applications 68(5), 1184–1194
75. Goberna MA, López, MA, Pastor J (1981): *Farkas–Minkowski systems in semi-infinite programming*. Applied Mathematics and Optimization 7(4), 295–308
76. Gowda MS, Teboulle M (1990): *A comparison of constraint qualifications in infinite-dimensional convex programming*. SIAM Journal on Control and Optimization 28(4), 925–935
77. Holmes RB (1975): *Geometric Functional Analysis and Its Applications*. Springer, New York Heidelberg Berlin
78. Hirriart-Urruty JB (1982): *ε-Subdifferential calculus*. In: Aubin JP, Vinter RB (eds) Convex Analysis and Optimization. Research Notes in Mathematics 57. Pitman, New York, 43–92
79. Hiriart-Urruty JB, Lemaréchal C (1993): *Convex Analysis and Minimization Algorithms I*. Springer, Berlin Heidelberg New York
80. Hu H (2005): *Characterizations of the strong basic constraint qualifications*. Mathematics of Operations Research 30(4), 956–965
81. Jahn J (1996): *Introduction to the Theory of Nonlinear Optimization*. Springer, Berlin Heidelberg New York Tokyo
82. Jeyakumar V (1997): *Asymptotic dual conditions characterizing optimality for infinite convex programs*. Journal of Optimization Theory and Applications 93(1), 153–165
83. Jeyakumar V (2006): *The strong conical hull intersection property for convex programming*. Mathematical Programming 106(1), 81–92
84. Jeyakumar V, Dinh N, Lee GM (2004): *A new closed cone constraint qualification for convex optimization*. Applied Mathematics Report AMR 04/8, University of New South Wales
85. Jeyakumar V, Lee GM, Dinh N (2003): *New sequential Lagrange multiplier conditions characterizing optimality without constraint qualification for convex programs*. SIAM Journal of Optimization 14(2), 534–547
86. Jeyakumar V, Song W, Dinh N, Lee GM (2005): *Stable strong duality in convex optimization*. Applied Mathematics Report AMR 05/22, University of New South Wales
87. Jeyakumar V, Wolkowicz H (1992): *Generalizations of Slater's constraint qualification for infinite convex programs*. Mathematical Programming 57(1), 85–101
88. Kinderlehrer D, Stampacchia G (1980): *An Introduction to Variational Inequalities and Their Applications*. Classics in Applied Mathematics 31. Society for Industrial and Applied Mathematics, Philadelphia
89. Köthe G (1960): *Topologische Lineare Räume I*. Die Grundlehren der Mathematischen Wissenschaften 107. Springer, Berlin Göttingen Heidelberg
90. Krauss E (1985): *A representation of arbitrary maximal monotone operators via subgradients of skew-symmetric saddle functions*. Nonlinear Analysis 9(12), 1381–1399
91. Krauss E (1985): *A representation of maximal monotone operators by saddle functions*. Revue Roumaine de Mathématiques Pures et Appliquées 30(10), 823–836
92. Kunen K, Rosenthal H (1982): *Martingale proofs of some geometrical results in Banach space theory*. Pacific Journal of Mathematics 100(1), 153–175

93. Limber MA, Goodrich RK (1993): *Quasi interiors, Lagrange multipliers, and L^p spectral estimation with lattice bounds*. Journal of Optimization Theory and Applications 78(1), 143–161
94. Luc DT (1989): *Theory of Vector Optimization*. Springer, Berlin Heidelberg New York
95. Magnanti TL (1974): *Fenchel and Lagrange duality are equivalent*. Mathematical Programming 7, 253–258
96. Marquez Alves M, Svaiter BF (2008): *Brønsted–Rockafellar property and maximality of monotone operators representable by convex functions in non-reflexive Banach spaces*. Journal of Convex Analysis 15(4), 693–706
97. Martinez-Legaz JE, Théra M (1996): *ε-Subdifferentials in terms of subdifferentials*. Set-Valued Analysis 4(4), 327–332
98. Martinez-Legaz JE, Théra M (2001): *A convex representation of maximal monotone operators*. Journal of Nonlinear and Convex Analysis 2(2), 243–247
99. Mordukhovich BS (2006): *Variational Analysis and Generalized Differentiation, I: Basic Theory*. Springer, Berlin
100. Mordukhovich BS (2006): *Variational Analysis and Generalized Differentiation, II: Applications*. Springer, Berlin
101. Morris P (1983): *Disappearance of extreme points*. Proceedings of the American Mathematical Society 88(2), 244–246
102. Ng KF, Song W (2003): *Fenchel duality in infinite-dimensional setting and its applications*. Nonlinear Analysis: Theory, Methods & Applications 55(7–8), 845–858
103. Pennanen T (2000): *Dualization of generalized equations of maximal monotone type*. SIAM Journal on Optimization 10(3), 809–835
104. Penot JP (2003): *Is convexity useful for the study of monotonicity?* In: Agarwal RP, O'Regan D (eds) Nonlinear Analysis and Its Applications, Kluwer, Dordrecht, 807–822
105. Penot JP (2004): *The relevance of convex analysis for the study of monotonicity*. Nonlinear Analysis: Theory, Methods & Applications 58(7–8), 855–871
106. Penot JP (2004): *A representation of maximal monotone operators by closed convex functions and its impact on calculus rules*. Comptes Rendus Mathématique, Académie des Sciences Paris 338(11), 853–858
107. Penot JP, Théra M (1982): *Semicontinuous mappings in general topology*. Archiv der Mathematik 38(2), 158–166
108. Penot JP, Zălinescu C (2005): *Some problems about the representation of monotone operators by convex functions*. The Australian & New Zealand Industrial and Applied Mathematics Journal 47(1), 1–20
109. Phelps RR (1961): *Extreme points of polar convex sets*. Proceedings of the American Mathematical Society 12(2), 291–296
110. Precupanu T (1984): *Closedness conditions for the optimality of a family of non-convex optimization problem*. Mathematische Operationsforschung und Statistik Series Optimization 15(3), 339–346
111. Revalski JP, Théra M (2002): *Enlargements and sums of monotone operators*. Nonlinear Analysis: Theory, Methods & Applications 48(4), 505–519
112. Rockafellar RT (1966): *Extension of Fenchel's duality theorem for convex functions*. Duke Mathematical Journal 33(1), 81–89
113. Rockafellar RT (1970): *Convex Analysis*. Princeton University Press, Princeton
114. Rockafellar RT (1970): *On the maximal monotonicity of subdifferential mappings*. Pacific Journal of Mathematics 33, 209–216
115. Rockafellar RT (1970): *On the maximality of sums of nonlinear monotone operators*. Transactions of the American Mathematical Society 149, 75–88
116. Rockafellar RT (1974): *Conjugate Duality and Optimization*. Regional Conference Series in Applied Mathematics 16. Society for Industrial and Applied Mathematics, Philadelphia
117. Rodrigues B (1990): *The Fenchel duality theorem in Fréchet spaces*. Optimization 21(1), 13–22

118. Rodrigues B, Simons S (1988): *Conjugate functions and subdifferentials in non-normed situations for operators with complete graphs*. Nonlinear Analysis: Theory, Methods & Applications 12(10), 1069–1078
119. Simons S (1998): *Minimax and Monotonicity*. Springer, Berlin Heidelberg New York
120. Simons S (2008): *From Hahn–Banach to Monotonicity*. Springer, Berlin Heidelberg
121. Simons S, Zălinescu C (2005): *Fenchel duality, Fitzpatrick functions and maximal monotonicity*. Journal of Nonlinear and Convex Analysis 6(1), 1–22
122. Strömberg T (1996): *The operation of infimal convolution*. Dissertationes Mathematicae 352, 1–58
123. Tiba D, Zălinescu C (2004): *On the necessity of some constraint qualification conditions in convex programming*. Journal of Convex Analysis 11(1), 95–110
124. Wanka G, Boţ RI (2002): *On the relations between different dual problems in convex mathematical programming*. In: Chamoni P, Leisten R, Martin A, Minnemann J, Stadtler H (eds) Operations Research Proceedings 2001. Springer, Heidelberg, 255–262
125. Zălinescu C (1983): *Duality for vectorial nonconvex optimization by convexification and applications*. Analele Ştiinţifice ale Universităţii Alexandru Ioan Cuza Iaşi Secţia I-a Matematică 29(3), 15–34
126. Zălinescu C (1987): *Solvability results for sublinear functions and operators*. Mathematical Methods of Operations Research 31(3), A79–A101
127. Zălinescu C (2002): *Convex Analysis in General Vector Spaces*. World Scientific, River Edge
128. Zălinescu C (2005): *A new proof of the maximal monotonicity of the sum using the Fitzpatrick function*. In: Giannessi F, Maugeri A (eds) Variational analysis and applications. Nonconvex Optimization and Its Applications 79. Kluwer, Dordrecht, 1159–1172
129. Zălinescu C (2007): *On the second conjugate of several convex functions in general normed vector spaces*. Journal of Global Optimization 40(1–3), 475–487
130. Zeidler E (1998): *Nonlinear Functional Analysis and Its Applications III. Variational Methods and Optimization*. Springer, Berlin

Index

Lightning Source UK Ltd.
Milton Keynes UK
05 March 2010

150932UK00001B/33/P

9 783642 048999